REAL AND CONVEX ANALYSIS

Real and Convex Analysis

Editor

Qing Jun Hou

scitus academics

Real and Convex Analysis

Edited by **Qing Jun Hou**

ISBN: 978-1-68117-566-9

Library of Congress Control Number: 2016942996

SCITUS Academics LLC,
www.scitusacademics.com
Box No. 4766, 616 Corporate Way,
Suite 2, Valley Cottage,
NY 10989

Printed in United States of America on Acid Free Paper

PREFACE

Real analysis is an area of mathematics that deals with sets and sequences of real numbers, as well as functions of one or more real variables. As one of the main branches of analysis, it can be seen as a subset of complex analysis, many results of the former being special cases of results in the latter. Real analysis deals with the real numbers and real-valued functions of a real variable. In particular, it deals with the analytic properties of real functions and sequences, including convergence and limits of sequences of real numbers, the calculus of the real numbers, and continuity, smoothness and related properties of real-valued functions. Convex analysis is devoted to the study of properties of convex functions and convex sets, often with applications in convex minimization, a subdomain of optimization theory. One of the fields of application of convex analysis is optimization, meaning, the search for maxima or minima of some functions, and for points at which such extrema are reached. Real-analysis is necessary for probability theory, which is the foundation for all of statistics, operations research, queuing theory, and the mathematical finance. Convex analysis is the mathematical foundation for convex optimization, having deep knowledge of real and convex analysis helps students and researchers apply its tools more effectively.

Real and Convex Analysis aims to provide a concise, accessible account of real and convex analysis and its applications and extensions, for a broad audience. It will be of valuable tool for professors, researchers, scientists and engineers. It can also be used for the advanced undergraduate level students.

TABLE OF CONTENTS

7

ADOMIAN DECOMPOSITION METHOD WITH GREEN'S FUNCTION FOR SOLVING TENTH-ORDER BOUNDARY VALUE PROBLEMS

Waleed Al-Hayani

Department of Mathematics, College of Computer Science and Mathematics, University of Mosul, Mosul, Iraq

ABSTRACT

In this paper, the Adomian decomposition method with Green's function (Standard Adomian and Modified Technique) is applied to solve linear and nonlinear tenth-order boundary value problems with boundary conditions defined at any order derivatives. The numerical results obtained with a small amount of computation are compared with the exact solutions to show the efficiency of the method. The results show that the decomposition method is of high accuracy, more convenient and efficient for solving high-order boundary value problems.

KEYWORDS

Adomian decomposition method, Adomian's polynomials, Tenth-order boundary value problems, Green's function

1. INTRODUCTION

In the beginning of the 1980's, Adomian [1] -[4] proposed a new and fruitful method (hereafter called the Adomian Decomposition Method or ADM) for solving linear and nonlinear (algebraic, differential, partial differential, integral, etc.) equations. It has been shown that this method yields a rapid convergence of the solutions series to linear and nonlinear deterministic and stochastic equations.

The lower-order boundary value problems have been vastly examined, analytically and numerically, in the literature. In contrast, higher-order boundary value problems have not been studied to the same extent that lower- order equations have been investigated. Nowadays, higher-order boundary value problems receive an increased interest due to the fact that they are noted in many mathematical physics applications. A class of characteristic- value problems of high order (as high as twenty four) are known to arise in hydrodynamic and hydromagnetic stability [5]. Tenth-order differential equations govern the physics of some hydrodynamic stability problems. When an infinite horizontal layer of fluid is heated from below, with the supposition that a uniform magnetic field is also applied across the fluid in the same direction as gravity and the fluid is subject to the action of rotation, instability sets in. When this instability sets in as ordinary convection, it is modelled by a tenth-order ordinary differential equation [5] .

Theorems which list the conditions for the existence and uniqueness of solutions of BVPs of higher order are thoroughly investigated in a book by Agarwal [6] . However, no numerical methods are contained in [6] for solving such problems.

Different numerical and semi analytical methods have been proposed by various authors to solve tenth-order boundary-value problems. A few of them are: Tenth degree spline method [7] , Modified Decomposition Method with the inverse operator (MDM) [8] , Differential Transform Method (DTM) [9] , Eleventh Degree Spline Method (EDSM) [10] , Non-Polynomial Spline Method (NPSM) [11] , Variational Iteration Technique (VIT) [12] and Homotopy Perturbation Method (HPM) [13] .

The main objective of this paper is to apply the Standard Adomian with Green's function (SAwGF) and Modified Technique with Green's function (MTwGF) to linear and nonlinear tenth-order boundary value problems with boundary conditions defined at any order derivatives.

2. ANALYSIS OF THE METHOD

Let us consider the general BVP of tenth-order

$$y^{(10)}(x)+g(x,y)=f(x), a\le x\le b \tag{1}$$

with boundary conditions

$$y^{(i)}(a)=\alpha_i, y^{(i)}(b)=\beta_i, i=0,1,2,3,4 \tag{2}$$

where $y=y(x)$, $g(x,y)$ is a linear or nonlinear function of y, and $f(x)$ are continuous functions defined in the interval $x\in[a,b]$ and α_i, β_i; $(i=0,1,2,3,4)$ are finite real constants.

Applying the decomposition method as in [1] -[4] , Equation (1) can be written as

$$Ly=f(x)-Ny,$$

where $L=\frac{d^{10}}{dx^{10}}$ is the linear operator and $Ny=g(x,y)$ is the nonlinear operator. Consequently,

$$y(x)=h(x)+\int_a^b G(x,\xi)f(\xi)\mathrm{d}\xi-\int_a^b G(x,\xi)Ny\mathrm{d}\xi, \tag{3}$$

where $h(x)$ is the solution of $Ly=0$ with the boundary conditions (2) and $G(x,\xi)$ is the Green's function [14] given by

$$G(x,\xi)=\begin{cases} g_2(x,\xi) & \text{if} \quad a\le x\le \xi\le b \\ g_1(x,\xi) & \text{if} \quad a\le \xi\le x\le b \end{cases}$$

The Adomian's technique consists of approximating the solution of (1) as an infinite series

$$y=\sum_{n=0}^{\infty} y_n, \tag{4}$$

and decomposing the nonlinear operator N as

$$Ny=\sum_{n=0}^{\infty} A_n, \tag{5}$$

where A_n are polynomials (called Adomian polynomials) of $y_0, y_1, \cdots, y_n$ [1] - [4] given by

$$A_n=\frac{1}{n!}\frac{\mathrm{d}^n}{\mathrm{d}\lambda^n}\left[N\left(\sum_{i=0}^{\infty}\lambda^i y_i\right)\right]_{\lambda=0}, n=0,1,2,\cdots.$$

The proofs of the convergence of the series $\sum_{n=0}^{\infty} y_n$ and $\sum_{n=0}^{\infty} A_n$ are given in [3] [15] - [19] . Substituting (4)

and (5) into (3) yields

$$\sum_{n=0}^{\infty} y_n = h(x)+\int_a^b G(x,\xi)f(\xi)\mathrm{d}\xi-\int_a^b G(x,\xi)\sum_{n=0}^{\infty} A_n \mathrm{d}\xi. \tag{6}$$

From (6), the iterates defined using the Standard Adomian Method are determined in the following recursive way:

$$y_0 = h(x)+\int_a^b G(x,\xi)f(\xi)\mathrm{d}\xi,$$

$$y_{n+1} = -\int_a^b G(x,\xi) A_n \mathrm{d}\xi, n = 0,1,2,\cdots$$

and the iterates defined using the Modified Technique [20] are determined in the following recursive way:

$$y_0 = h(x),$$

$$y_1 = \int_a^b G(x,\xi) f(\xi) \mathrm{d}\xi - \int_a^b G(x,\xi) A_0 \mathrm{d}\xi,$$

$$y_{n+2} = -\int_a^b G(x,\xi) A_{n+1} \mathrm{d}\xi, n = 0,1,2,\cdots$$

Thus all components of y can be calculated once the A_n are given. We then define the n-term approximant to the solution y by $\phi_n[y] = \sum_{i=0}^{n-1} y_i$ with $\lim_{n\to\infty} \phi_n[y] = y$.

3. APPLICATIONS AND NUMERICAL RESULTS

In this section, the ADM with the Green's function (Standard Adomian and Modified Technique) for solving linear and nonlinear tenth-order boundary value problems is illustrated in the following examples. To show the high accuracy of the solution results compared with the exact solution, we give the maximum absolute error and the maximum residual error. The computations associated with the examples were performed using a Maple 13 package with a precision of 40 dígits.

3.1. Example 1

Consider the following linear BVP of tenth-order [9] - [12] :

$$y^{(10)}(x) - xy(x) = -\left(89 + 21x + x^2 - x^3\right)e^x, -1 \le x \le 1 \qquad (7)$$

with boundary conditions

$$\left.\begin{aligned} &y(-1)=0, y(1)=0, \\ &y^{(1)}(-1)=2e^{-1}, y^{(1)}(1)=-2e, \\ &y^{(2)}(-1)=2e^{-1}, y^{(2)}(1)=-6e, \\ &y^{(3)}(-1)=0, y^{(3)}(1)=-12e, \\ &y^{(4)}(-1)=-4e^{-1}, y^{(4)}(1)=-20e. \end{aligned}\right\} \tag{8}$$

The exact solution of (7), (8) is

$$y_{Exact}(x)=\left(1-x^2\right)e^x.$$

Applying the decomposition method, Equation (7) can be written as

$$Ly=-\left(89+21x+x^2-x^3\right)e^x+xy(x),$$

where $L=\frac{d^{10}}{dx^{10}}$ is the linear operator. Consequently,

$$\begin{aligned} y=h(x)&-\int_{-1}^{1}G(x,\xi)\left(89+21\xi+\xi^2-\xi^3\right)e^{\xi}d\xi \\ &+\int_{-1}^{1}G(x,\xi)\xi y(\xi)d\xi, \end{aligned} \tag{9}$$

where $h(x)$ is the solution of $Ly=0$ with the boundary conditions (8) given by

$$\begin{aligned} h(x)=&\frac{1}{96e}\left\{\left(5e^2-37\right)x^9-\left(e^2-7\right)x^8-\left(26e^2-190\right)x^7+\left(4e^2-40\right)x^6\right\} \\ &+\frac{1}{96e}\left\{\left(48e^2-396\right)x^5-\left(30e^2-102\right)x^4-\left(86e^2-418\right)x^3\right\} \\ &+\frac{1}{96e}\left\{\left(4e^2-160\right)x^2+\left(59e^2-175\right)x+23e^2+91\right\} \end{aligned}$$

and $G(x,\xi)$ is the Green's function given by

$$G(x,\xi)=\begin{cases} g_2(x,\xi) & \text{if} \quad -1\le x\le \xi \le 1 \\ g_1(x,\xi) & \text{if} \quad -1\le \xi \le x\le 1 \end{cases}$$

where

$$\begin{aligned} g_1(x,\xi) = & \left(\frac{1}{2654208}x^9 - \frac{1}{516096}x^7 + \frac{1}{245760}x^5 - \frac{1}{221184}x^3 + \frac{1}{294912}x - \frac{1}{725760}\right)\xi^9 \\ & + \left(\frac{1}{2064384}x^8 - \frac{1}{368640}x^6 + \frac{1}{147456}x^4 - \frac{1}{73728}x^2 + \frac{1}{80640}x - \frac{1}{294912}\right)\xi^8 \\ & - \left(\frac{1}{516096}x^9 - \frac{1}{92160}x^7 + \frac{1}{36864}x^5 - \frac{1}{18432}x^3 + \frac{1}{20160}x^2 - \frac{1}{73728}x\right)\xi^7 \\ & - \left(\frac{1}{368640}x^8 - \frac{1}{55296}x^6 + \frac{1}{12288}x^4 - \frac{1}{8640}x^3 + \frac{1}{18432}x^2 - \frac{1}{221184}\right)\xi^6 \\ & + \left(\frac{1}{245760}x^9 - \frac{1}{36864}x^7 + \frac{1}{8192}x^5 - \frac{1}{5760}x^4 + \frac{1}{12288}x^3 - \frac{1}{147456}x\right)\xi^5 \\ & + \left(\frac{1}{147456}x^8 - \frac{1}{12288}x^6 + \frac{1}{5760}x^5 - \frac{1}{8192}x^4 + \frac{1}{36864}x^2 - \frac{1}{245760}\right)\xi^4 \\ & - \left(\frac{1}{221184}x^9 - \frac{1}{18432}x^7 + \frac{1}{8640}x^6 - \frac{1}{12288}x^5 + \frac{1}{55296}x^3 - \frac{1}{368640}x\right)\xi^3 \\ & - \left(\frac{1}{73728}x^8 - \frac{1}{20160}x^7 + \frac{1}{18432}x^6 - \frac{1}{36864}x^4 + \frac{1}{92160}x^2 - \frac{1}{516096}\right)\xi^2 \\ & + \left(\frac{1}{294912}x^9 - \frac{1}{80640}x^8 + \frac{1}{73728}x^7 - \frac{1}{147456}x^5 + \frac{1}{368640}x^3 - \frac{1}{2064384}x\right)\xi \\ & + \frac{1}{725760}x^9 - \frac{1}{294912}x^8 + \frac{1}{221184}x^6 - \frac{1}{245760}x^4 + \frac{1}{516096}x^2 - \frac{1}{2654208} \end{aligned}$$

$$
\begin{aligned}
g_2(x,\xi)=&\left(\frac{1}{2654208}\xi^9-\frac{1}{516096}\xi^7+\frac{1}{245760}\xi^5-\frac{1}{221184}\xi^3+\frac{1}{294912}\xi-\frac{1}{725760}\right)x^9\\
&+\left(\frac{1}{2064384}\xi^8-\frac{1}{368640}\xi^6+\frac{1}{147456}\xi^4-\frac{1}{73728}\xi^2+\frac{1}{80640}\xi-\frac{1}{294912}\right)x^8\\
&-\left(\frac{1}{516096}\xi^9-\frac{1}{92160}\xi^7+\frac{1}{36864}\xi^5-\frac{1}{18432}\xi^3+\frac{1}{20160}\xi^2-\frac{1}{73728}\xi\right)x^7\\
&-\left(\frac{1}{368640}\xi^8-\frac{1}{55296}\xi^6+\frac{1}{12288}\xi^4-\frac{1}{8640}\xi^3+\frac{1}{18432}\xi^2-\frac{1}{221184}\right)x^6\\
&+\left(\frac{1}{245760}\xi^9-\frac{1}{36864}\xi^7+\frac{1}{8192}\xi^5-\frac{1}{5760}\xi^4+\frac{1}{12288}\xi^3-\frac{1}{147456}\xi\right)x^5\\
&+\left(\frac{1}{147456}\xi^8-\frac{1}{12288}\xi^6+\frac{1}{5760}\xi^5-\frac{1}{8192}\xi^4+\frac{1}{36864}\xi^2-\frac{1}{245760}\right)x^4\\
&-\left(\frac{1}{221184}\xi^9-\frac{1}{18432}\xi^7+\frac{1}{8640}\xi^6-\frac{1}{12288}\xi^5+\frac{1}{55296}\xi^3-\frac{1}{368640}\xi\right)x^3\\
&-\left(\frac{1}{73728}\xi^8-\frac{1}{20160}\xi^7+\frac{1}{18432}\xi^6-\frac{1}{36864}\xi^4+\frac{1}{92160}\xi^2-\frac{1}{516096}\right)x^2\\
&+\left(\frac{1}{294912}\xi^9-\frac{1}{80640}\xi^8+\frac{1}{73728}\xi^7-\frac{1}{147456}\xi^5+\frac{1}{368640}\xi^3-\frac{1}{2064384}\xi\right)x\\
&+\frac{1}{725760}\xi^9-\frac{1}{294912}\xi^8+\frac{1}{221184}\xi^6-\frac{1}{245760}\xi^4+\frac{1}{516096}\xi^2-\frac{1}{2654208}.
\end{aligned}
$$

Substituting (4) in (9), the iterates defined using the Standard Adomian Method are determined in the following recursive way:

$$y_0=h(x)-\int_{-1}^{1}G(x,\xi)\left(89+21\xi+\xi^2-\xi^3\right)e^{\xi}d\xi,$$

$$y_{n+1}=\int_{-1}^{1}G(x,\xi)\xi y_n(\xi)d\xi, n=0,1,2,\cdots$$

and the iterates defined using the Modified Technique [20] are determined in the following recursive way:

$$y_0=h(x),$$

$$y_1=-\int_{-1}^{1}G(x,\xi)\left(89+21\xi+\xi^2-\xi^3\right)e^{\xi}d\xi+\int_{-1}^{1}G(x,\xi)\xi y_0(\xi)d\xi,$$

$$y_{n+2} = \int_{-1}^{1} G(x,\xi)\xi y_{n+1}(\xi)\mathrm{d}\xi, n = 0,1,2,\cdots.$$

In Table 1, we list the maximum errors obtained by SAwGF and MTwGF with the exact solution. Comparing them with the DTM [9] , EDSM [10] , NPSM [11] and VIT [12] results, we notice that the result obtained by the present method (SAwGF) is very superior (lower error combined with less number of iterations) to that obtained by the other mentioned methods. Table 2 reproduces the maximum residual error of the SAwGF and MTwGF for ϕ_n.

3.2. Example 2

Consider the following linear BVP of tenth-order [9] - [12]:

$$y^{(10)}(x) + y(x) = -10(2x\sin x - 9\cos x), -1 \le x \le 1 \tag{10}$$

with boundary conditions

Table 1. Comparison of maximum errors for example 1.

	SAwGF	MTwGF	DTM [9]	EDSM [10]	NPSM [11]	VIT [12]
	$n=2$	$n=2$	$n=13$			$n=15$
$\|\cdot\|_\infty$	1.99E-16	6.69E-14	2.70E-08	3.28E-06	4.72E-06	1.97E-06
$\|\cdot\|_2$	1.41E-16	4.56E-14	-	-	-	-

Table 2. Maximum residual error for example 1.

	SAwGF		MTwGF	
n	$\|E_r(\phi_n)\|_\infty$	$\|E_r(\phi_n)\|_2$	$\|E_r(\phi_n)\|_\infty$	$\|E_r(\phi_n)\|_2$
2	4.51E-09	2.94E-09	5.02E-06	3.98E-06
3	4.64E-17	3.11E-17	2.23E-14	1.48E-14
4	4.17E-25	2.71E-25	4.63E-22	3.68E-22

$$\left.\begin{aligned}&y(-1)=0, y(1)=0,\\&y^{(1)}(-1)=-2\cos(1), y^{(1)}(1)=2\cos(1),\\&y^{(2)}(-1)=2\cos(1)-4\sin(1), y^{(2)}(1)=2\cos(1)-4\sin(1),\\&y^{(3)}(-1)=6\cos(1)+6\sin(1), y^{(3)}(1)=-6\cos(1)-6\sin(1),\\&y^{(4)}(-1)=-12\cos(1)+8\sin(1), y^{(4)}(1)=-12\cos(1)+8\sin(1).\end{aligned}\right\} \tag{11}$$

The exact solution of (10), (11) is

$$y_{\text{Exact}}(x)=(x^2-1)\cos x.$$

Applying the decomposition method, Equation (10) can be written as

$$Ly=-10(2x\sin x-9\cos x)-y(x),$$

where $L=\frac{d^{10}}{dx^{10}}$ is the linear operator. Consequently,

$$y=h(x)-10\int_{-1}^{1}G(x,\xi)(2\xi\sin\xi-9\cos\xi)\mathrm{d}\xi-\int_{-1}^{1}G(x,\xi)y(\xi)\mathrm{d}\xi, \tag{12}$$

where $h(x)$ is the solution of $Ly=0$ with the boundary conditions (11) given by

$$h(x)=-\left(\frac{1}{24}\sin 1-\frac{1}{16}\cos 1\right)x^8+\left(\frac{7}{24}\sin 1-\frac{3}{8}\cos 1\right)x^6-\left(\frac{9}{8}\sin 1-\frac{3}{4}\cos 1\right)x^4$$
$$+\left(\frac{37}{24}\sin 1+\frac{3}{8}\cos 1\right)x^2-\frac{2}{3}\sin 1-\frac{13}{16}\cos 1$$

and $G(x,\xi)$ is the Green's function given previously in example 1. Substituting (4) in (12), the iterates defined using the Standard Adomian Method are determined in the following recursive way:

$$y_0=h(x)-10\int_{-1}^{1}G(x,\xi)(2\xi\sin\xi-9\cos\xi)\mathrm{d}\xi,$$

$$y_{n+1} = -\int_{-1}^{1} G(x,\xi)\, y_n(\xi)\, \mathrm{d}\xi, n = 0,1,2,\cdots$$

and the iterates defined using the Modified Technique [20] are determined in the following recursive way:

$$y_0 = h(x),$$

$$y_1 = -10\int_{-1}^{1} G(x,\xi)(2\xi \sin\xi - 9\cos\xi)\,\mathrm{d}\xi - \int_{-1}^{1} G(x,\xi)\, y_0(\xi)\,\mathrm{d}\xi,$$

$$y_{n+2} = -\int_{-1}^{1} G(x,\xi)\, y_{n+1}(\xi)\,\mathrm{d}\xi, n = 0,1,2,\cdots.$$

In Table 3, we present the maximum errors obtained by SAwGF and MTwGF with the exact solution. Comparing them with the DTM [9] , EDSM [10] , NPSM [11] and VIT [12] results, it can be noticed that the result obtained by the present method (SAwGF) is very superior to that obtained by the other mentioned methods. Table 4 exhibits the maximum residual error of the SAwGF and MTwGF for ϕ_n.

3.3. Example 3

Consider the following linear BVP of tenth-order [10] :

$$y^{(10)}(x) - (x^2 - 2x)\, y(x) = -(x-1)^3 \sin x + 10\cos x, -1 \le x \le 1 \tag{13}$$

with boundary conditions

$$\left.\begin{aligned}
&y(-1) = 2\sin(1), y(1) = 0,\\
&y^{(1)}(-1) = -2\cos(1) - \sin(1), y^{(1)}(1) = \sin(1),\\
&y^{(2)}(-1) = 2\cos(1) - 2\sin(1), y^{(2)}(1) = 2\cos(1),\\
&y^{(3)}(-1) = 2\cos(1) + 3\sin(1), y^{(3)}(1) = -3\sin(1),\\
&y^{(4)}(-1) = -4\cos(1) + 2\sin(1), y^{(4)}(1) = -4\cos(1).
\end{aligned}\right\} \tag{14}$$

The exact solution of (13), (14) is

$$y_{\text{Exact}}(x)=(x-1)\sin x.$$

Applying the decomposition method, Equation (13) can be written as

$$Ly=-(x-1)^3\sin x+10\cos x+\left(x^2-2x\right)y(x),$$

where $L=\frac{d^{10}}{dx^{10}}$ is the linear operator. Consequently,

$$y=h(x)+\int_{-1}^{1}G(x,\xi)\left[-(\xi-1)^3\sin\xi+10\cos\xi\right]d\xi+\int_{-1}^{1}G(x,\xi)\left(\xi^2-2\xi\right)y(\xi)d\xi,$$

(15)

where $h(x)$ is the solution of $Ly=0$ with the boundary conditions (14) given by

$$\begin{aligned}h(x)=&-\left(\frac{61}{384}\sin 1-\frac{95}{384}\cos 1\right)x^9-\left(\frac{11}{384}\sin 1-\frac{17}{384}\cos 1\right)x^8+\left(\frac{79}{96}\sin 1-\frac{41}{32}\cos 1\right)x^7\\&+\left(\frac{17}{96}\sin 1-\frac{25}{96}\cos 1\right)x^6-\left(\frac{113}{64}\sin 1-\frac{175}{64}\cos 1\right)x^5-\left(\frac{39}{64}\sin 1-\frac{41}{64}\cos 1\right)x^4\\&+\left(\frac{211}{96}\sin 1-\frac{299}{96}\cos 1\right)x^3+\left(\frac{125}{96}\sin 1-\frac{17}{96}\cos 1\right)x^2-\left(\frac{805}{384}\sin 1-\frac{181}{128}\cos 1\right)x\\&+\frac{61}{384}\sin 1-\frac{95}{384}\cos 1\end{aligned}$$

and $G(x,\xi)$ is the Green's function given previously in example 1. Substituting (4) in (15), the iterates defined using the Standard Adomian Method are determined in the following recursive way:

$$y_0=h(x)+\int_{-1}^{1}G(x,\xi)\left[-(\xi-1)^3\sin\xi+10\cos\xi\right]d\xi,$$

$$y_{n+1}=\int_{-1}^{1}G(x,\xi)\left(\xi^2-2\xi\right)y_n(\xi)d\xi,n=0,1,2,\cdots$$

and the iterates defined using the Modified Technique [20] are determined in the following recursive way:

$$y_0 = h(x),$$

$$y_1 = \int_{-1}^{1} G(x,\xi)\left[-(\xi-1)^3 \sin\xi + 10\cos\xi\right]\mathrm{d}\xi + \int_{-1}^{1} G(x,\xi)\left(\xi^2 - 2\xi\right)y_0(\xi)\mathrm{d}\xi,$$

$$y_{n+2} = \int_{-1}^{1} G(x,\xi)\left(\xi^2 - 2\xi\right)y_{n+1}(\xi)\mathrm{d}\xi, n = 0,1,2,\cdots.$$

Table 3. Comparison of maximum errors for example 2.

	SAwGF	MTwGF	DTM [9]	EDSM [10]	NPSM [11]	VIT [12]
	$n=2$	$n=2$	$n=10$			$n=18$
$\|\cdot\|_\infty$	4.98E-14	4.80E-12	1.12E-06	8.85E-08	4.67E-07	4.24E-07
$\|\cdot\|_2$	3.66E-14	3.52E-12	-	-	-	-

Table 4. Maximum residual error for example 2.

	SAwGF		MTwGF	
n	$\|E_r(\phi_n)\|_\infty$	$\|E_r(\phi_n)\|_2$	$\|E_r(\phi_n)\|_\infty$	$\|E_r(\phi_n)\|_2$
2	2.46E-07	1.81E-07	2.37E-05	1.74E-05
3	4.98E-14	3.66E-14	4.80E-12	3.52E-12
4	1.01E-20	7.38E-21	9.69E-19	7.11E-19

In Table 5, we give the maximum errors obtained by SAwGF and MTwGF with the exact solution. Comparing them with the EDSM [10] results, it can be seen easily that the result obtained by the present method (SAwGF) is very superior to that obtained by the other mentioned method. Table 6reproduces the maximum residual error of the SAwGF and MTwGF for ϕ_n.

3.4. Example 4

Finally, we consider the following nonlinear BVP of tenth-order [8] [13]:

$$y^{(10)}(x) = e^{-x} y^2(x), 0 < x < 1 \tag{16}$$

with boundary conditions

$$y^{(2i)}(0) = 1, y^{(2i)}(1) = e, i = 0,1,2,3,4. \tag{17}$$

The exact solution of (16), (17) is

$$y_{\text{Exact}}(x) = e^x.$$

Table 5. Comparison of maximum errors for example 3.

	SAwGF		MTwGF		EDSM [10]
	$n=2$	$n=3$	$n=2$	$n=3$	$\mu=0, k=56$
$\|\cdot\|_\infty$	7.05E-16	2.29E-23	2.75E-14	1.16E-21	3.73E-08
$\|\cdot\|_2$	2.56E-16	1.05E-23	1.86E-14	8.60E-22	-

Table 6. Maximum residual error for example 3.

	SAwGF		MTwGF	
n	$\|E_r(\varphi_n)\|_\infty$	$\|E_r(\varphi_n)\|_2$	$\|E_r(\varphi_n)\|_\infty$	$\|E_r(\varphi_n)\|_2$
2	2.13E-08	1.47E-08	1.13E-06	8.17E-07
3	4.16E-16	2.56E-16	1.79E-14	1.06E-14
4	1.17E-23	7.77E-24	5.74E-22	3.88E-22

Applying the decomposition method, Equation (16) can be written as

$$Ly = e^{-x} Ny,$$

where $L = \frac{d^{10}}{dx^{10}}$ is the linear operator and $Ny = y^2$ is the nonlinear operator. Consequently,

$$y = h(x) + \int_0^1 G(x,\xi) e^{-\xi} Ny(\xi) d\xi, \tag{18}$$

where $h(x)$ is the solution of $Ly = 0$ with the boundary conditions (17) given by

$$h(x) = \frac{e-1}{9!}x^9 - \frac{1}{8!}x^8 + \frac{5e-8}{30240}x^7 + \frac{1}{6!}x^6 + \frac{307e-472}{43200}x^5 + \frac{1}{4!}x^4 + \frac{12863e-19856}{90720}x^3 + \frac{1}{2}x^2 + \frac{171549e-264704}{201600}x + 1$$

and $G(x,\xi)$ is the Green's function given by

$$G(x,\xi) = \begin{cases} g_2(x,\xi) & \text{if} \quad 0 \le x \le \xi \le 1 \\ g_1(x,\xi) & \text{if} \quad 0 \le \xi \le x \le 1 \end{cases}$$

where

$$\begin{aligned} g_1(x,\xi) = &\left(\frac{1}{362880}x - \frac{1}{362880}\right)\xi^9 \\ &+ \left(\frac{1}{30240}x^3 - \frac{1}{10080}x^2 + \frac{1}{15120}x\right)\xi^7 \\ &+ \left(\frac{1}{14400}x^5 - \frac{1}{2880}x^4 + \frac{1}{2160}x^3 - \frac{1}{5400}x\right)\xi^5 \\ &+ \left(\frac{1}{30240}x^7 - \frac{1}{4320}x^6 + \frac{1}{2160}x^5 - \frac{1}{1620}x^3 + \frac{1}{2835}x\right)\xi^3 \\ &+ \left(\frac{1}{362880}x^9 - \frac{1}{40320}x^8 + \frac{1}{15120}x^7 - \frac{1}{5400}x^5 + \frac{1}{2835}x^3 - \frac{1}{4725}x\right) \end{aligned}$$

$$\begin{aligned} g_2(x,\xi) = &\left(\frac{1}{362880}\xi - \frac{1}{362880}\right)x^9 \\ &+ \left(\frac{1}{30240}\xi^3 - \frac{1}{10080}\xi^2 + \frac{1}{15120}\xi\right)x^7 \\ &+ \left(\frac{1}{14400}\xi^5 - \frac{1}{2880}\xi^4 + \frac{1}{2160}\xi^3 - \frac{1}{5400}\xi\right)x^5 \\ &+ \left(\frac{1}{30240}\xi^7 - \frac{1}{4320}\xi^6 + \frac{1}{2160}\xi^5 - \frac{1}{1620}\xi^3 + \frac{1}{2835}\xi\right)x^3 \\ &+ \left(\frac{1}{362880}\xi^9 - \frac{1}{40320}\xi^8 + \frac{1}{15120}\xi^7 - \frac{1}{5400}\xi^5 + \frac{1}{2835}\xi^3 - \frac{1}{4725}\xi\right)x. \end{aligned}$$

Substituting (4) and (5) in (18), the iterates defined using the Standard Adomian Method are determined in the following recursive way:

$$y_{n+1} = \int_0^1 G(x,\xi)e^{-\xi} A_n d\xi, n = 0,1,2,\cdots$$

For the nonlinear term $Ny = y^2 = \sum_{n=0}^{\infty} A_n$ the corresponding Adomian polynomials are:

$$\begin{aligned} A_0 &= y_0^2, \\ A_1 &= 2y_0y_1, \\ A_2 &= 2y_0y_2 + y_1^2, \\ &\vdots \\ A_n &= \sum_{i=0}^{n} y_i y_{n-i}, n \geq i, n = 0,1,2,\cdots. \end{aligned}$$

Table 7. Comparison of maximum errors for example 4.

	SAwGF		MDM [8]	HPM [13]
	$n=2$	$n=3$	$n=4$	
$\|\cdot\|_\infty$	4.90E-10	1.34E-14	4.58E-06	1.45E-05
$\|\cdot\|_2$	3.47E-10	9.48E-15	-	-

Table 8. Maximum residual error for example 4.

n	$\|E_r(\phi_n)\|_\infty$	$\|E_r(\phi_n)\|_2$
2	4.59E-05	3.25E-05
3	1.30E-09	8.89E-10
4	4.04E-14	2.69E-14

In Table 7, we list the maximum errors obtained by SAwGF with the exact solution. Comparing it with the MDM with the inverse operator [8] and HPM [13] results, it can be noticed that the result obtained by the present method (SAwGF) is very superior to that obtained by the other two mentioned methods. Table 8 exhibits the maximum residual error of the SAwGF for ϕ_n.

4. CONCLUSION

Waleed Al-Hayani (Standard Adomian and Modified Technique) has been applied for solving linear and nonlinear tenth-order boundary value problems with boundary conditions defined at any order derivatives. Comparison of the results obtained by the present method with those obtained by the Tenth degree spline method, Modified decomposition method with the inverse operator, Differential transform method, Eleventh degree spline method, Non-polynomial spline method, Variational iteration technique and Homotopy perturbation method has revealed that the present method is superior because of the lower error and fewer required iterations. It has been shown that error is monotonically reduced with the increment of the integer n.

REFERENCES

1. Adomian, G. (1983) Stochastic Systems. Academic Press, New York.
2. Adomian, G. (1986) Nonlinear Stochastic Operator Equations. Academic Press, New York.
3. Adomian, G. (1989) Nonlinear Stochastic Systems Theory and Applications to Physics. Kluwer Academic Publishers, Dordrecht
4. Adomian, G. (1994) Solving Frontier Problems of Physics: The Decomposition Method. Kluwer Academic Publishers, Dordrecht.
5. Chandrasekhar, S. (1961) Hydrodynamic Hydromagnetic Stability. Clarendon Press, Oxford. (Reprinted: Dover Books, New York, 1981)
6. Agarwal, R.P. (1986) Boundary Value Problems for Higher-Order Differential Equations. World Scientific, Singapore.

7. Siddiqi, S.S. and Twizell, E.H. (1998) Spline Solutions of Linear Tenth-Order Boundary-Value Problems. International Journal of Computer Mathematics, 68, 345-362.

8. Wazwaz, A.W. (2000) Approximate Solutions to Boundary Value Problems of Higher Order by the Modified Decomposition Method. Computers and Mathematics with Applications, 40, 679-691.

9. Erturk, V.S. and Momani, S. (2007) A Reliable Algorithm for Solving Tenth-Order Boundary Value Problems. Numerical Algorithms, 44, 147-158.

10. Siddiqi, S.S. and Akram, G. (2007) Solutions of Tenth-Order Boundary Value Problems using Eleventh Degree Spline. Applied Mathematics and Computation, 185, 115-127.

11. Siddiqi, S.S. and Akram, G. (2007) Solution of 10th-Order Boundary Value Problems Using Non-Polynomial Spline Technique. Applied Mathematics and Computation, 190, 641-651.

12. Siddiqi, S.S., Akram, G. and Zaheer, S. (2009) Solution of Tenth Order Boundary Value Problems Using Variational Iteration Technique. European Journal of Scientific Research, 30, 326-347.

13. Ravi Kanth, A.S.V. and Aruna, K. (2009) He's Homotopy Perturbation Method for Solving Higher-Order Boundary Value Problems. Chaos, Solitons and Fractals, 41, 1905-1909.

14. Stakgold, I. (1998) Green's Functions and Boundary Value Problems. John Wiley & Sons, Inc., Hoboken.

15. Abbaoui, K. and Cherruault, Y. (1994) Convergence of Adomian's Method Applied to Differential Equations. Mathematical and Computer Modelling, 28, 103-109.

16. Abbaoui, K. and Cherruault, Y. (1995) New Ideas for Proving Convergence of Decomposition Methods. Computers and Mathematics with Applications, 29, 103-108.

17. Abbaoui, K. and Cherruault, Y. (1994) Convergence of Adomian's Method Applied to Nonlinear Equations. Mathematical and Computer Modelling, 20, 69-73.

18. Cherruault, Y. and Adomian, G. (1993) Decomposition Methods: A New Proof of Convergence. Mathematical and Computer Modelling, 18, 103-106.

19. Guellal, S. and Cherruault, Y. (1994) Practical Formula for Calculation of Adomian's Polynomials and Application to the Convergence of the Decomposition Method. International Journal Bio-Medical Computing, 36, 223-228.

20. Wazwaz, A.W. (1999) A Reliable Modification of Adomian Decomposition Method. Applied Mathematics and Computation, 102, 77-86.

2

FOURIER COEFFICIENTS OF A CLASS OF ETA QUOTIENTS OF WEIGHT 16 WITH LEVEL 12

Barış Kendirli

Istanbul Kultur University, Istanbul, Turkey

ABSTRACT

Recently, Williams [1] and then Yao, Xia and Jin [2] discovered explicit formulas for the coef- ficients of the Fourier series expansions of a class of eta quotients. Williams expressed all coefficients of 126 eta quotients in terms of $\sigma(n), \sigma\left(\frac{n}{2}\right), \sigma\left(\frac{n}{3}\right)$ and $\sigma\left(\frac{n}{6}\right)$ and Yao, Xia and Jin, following the method of proof of Williams, expressed only even coefficients of 104 eta quotients in terms of $\sigma_3(n), \sigma_3\left(\frac{n}{2}\right), \sigma_3\left(\frac{n}{3}\right)$ and $\sigma_3\left(\frac{n}{6}\right)$. Here, by using the method of proof of Williams, we will express the even Fourier coefficients of 360 eta quotients i.e., the Fourier coefficients of the sum, f(q) + f(−q), of 360 eta quotients in terms of $\sigma_{15}(n), \sigma_{15}\left(\frac{n}{2}\right), \sigma_{15}\left(\frac{n}{3}\right), \sigma_{15}\left(\frac{n}{4}\right), \sigma_{15}\left(\frac{n}{6}\right)$ and $\sigma_{15}\left(\frac{n}{12}\right)$.

KEYWORDS

Dedekind Eta Function, Eta Quotients, Fourier Series

The divisor function $\sigma_i(n)$ is defined for a positive integer i by

$$\sigma_i(n) := \sum_{d \text{ positive integer}, d|n} d^i, \text{if } n \text{ is a positive integer, and}$$

$$\sigma_i(n) := 0 \text{ if } n \text{ is not a positive integer.} \tag{1}$$

The Dedekind eta function is defined by

$$\eta(z) := q^{1/24} \prod_{n=1}^{\infty} (1 - q^n), \tag{2}$$

where

$$q := e^{2\pi iz}, z \in H = \{x + iy : y > 0\} \tag{3}$$

And an eta quotient of level n is defined by

$$f(z) := \prod_{m|n} \eta(mz)^{a_m}, n \in \mathbb{N}, a_m \in Z, a_n \neq 0. \tag{4}$$

It is interesting and important to determine explicit formulas of the Fourier coefficients of eta quotients since they are the building blocks of modular forms of level n and weight k. The book of Köhler [3] (Chapter 3, p. 39) describes such expansions by means of Hecke Theta series and develops algorithms for the determination of suitable eta quotients. One can find more information in [4] -[8] . I have determined the Fourier coefficients of the theta series associated to some quadratic forms, see [9] - [14] .

Recently, Williams, see [1] discovered explicit formulas for the coefficients of Fourier series expansions of a class of 126 eta quotients in terms of $\sigma(n), \sigma\left(\frac{n}{2}\right), \sigma\left(\frac{n}{3}\right)$ and $\sigma\left(\frac{n}{6}\right)$. One example is as follows:

$$\frac{\eta^2(2z)\eta^4(4z)\eta^6(6z)}{\eta^2(z)\eta^2(3z)\eta^4(12z)}$$

gives the expansion found by Williams.

Then Yao, Xia and Jin [2] expressed the even Fourier coefficients of 104 eta quotients in terms of $\sigma_3(n), \sigma_3\left(\frac{n}{2}\right), \sigma_3\left(\frac{n}{3}\right)$ and $\sigma_3\left(\frac{n}{6}\right)$. One example is as follows:

$$\frac{\eta^{25}(2z)\eta^{4}(3z)}{\eta^{12}(z)\eta^{5}(4z)\eta^{3}(6z)\eta(12z)},$$

where the even coefficients are obtained. Motivated by these two results, we find that we can express the even Fourier coefficients of 360 eta quotients in terms of $\sigma_{15}(n), \sigma_{15}\left(\frac{n}{2}\right), \sigma_{15}\left(\frac{n}{3}\right), \sigma_{15}\left(\frac{n}{4}\right), \sigma_{15}\left(\frac{n}{6}\right)$ and $\sigma_{15}\left(\frac{n}{12}\right)$, see Table 2. One example is as follows:

$$\frac{\eta^{12}(4z)\eta^{10}(6z)\eta^{4}(12z)}{\eta^{6}(2z)}.$$

We see that the odd Fourier coefficients of 875 eta quotients are zero and even coefficients can be expressed by simple formula. Let

$$f_1 = \frac{\eta^{18}(4z)\eta^{16}(6z)\eta^{10}(12z)}{\eta^{12}(2z)},$$

$$f_2 = \frac{\eta^{20}(4z)\eta^{2}(6z)\eta^{20}(12z)}{\eta^{10}(2z)},$$

$$f_3 = \frac{\eta^{15}(4z)\eta^{7}(6z)\eta^{19}(12z)}{\eta^{9}(2z)},$$

$$f_4 = \frac{\eta^{10}(2z)\eta^{12}(4z)\eta^{18}(6z)}{\eta^{8}(12z)},$$

$$f_5 = \frac{\eta^{5}(4z)\eta^{17}(6z)\eta^{17}(12z)}{\eta^{7}(2z)},$$

$$f_6 = \frac{\eta^{17}(4z)\eta^{17}(6z)\eta^{5}(12z)}{\eta^{7}(2z)},$$

$$f_7 = \frac{\eta^{19}(4z)\eta^{15}(6z)\eta^{15}(12z)}{\eta^{17}(2z)},$$

$$f_8 = \frac{\eta^{17}(2z)\eta^{17}(4z)\eta^{5}(12z)}{\eta^{7}(6z)},$$

$$f_9 = \frac{\eta^{20}(2z)\eta^{14}(4z)\eta^{14}(12z)}{\eta^{16}(6z)},$$

$$f_{10} = \frac{\eta^{16}(4z)\eta^{18}(6z)}{\eta^{2}(2z)},$$

$$f_{11} = \frac{\eta^{20}(2z)\eta^{14}(4z)\eta^{8}(6z)}{\eta^{10}(12z)},$$

$$f_{12} = \eta^{20}(2z)\eta^{2}(4z)\eta^{8}(6z)\eta^{2}(12z),$$

$$f_{13} = \frac{\eta^{2}(2z)\eta^{14}(6z)\eta^{20}(12z)}{\eta^{4}(4z)},$$

$$f_{14} = \frac{\eta^{16}(2z)\eta^{10}(4z)\eta^{18}(12z)}{\eta^{12}(6z)},$$

$$f_{15} = \frac{\eta^{19}(4z)\eta^{9}(6z)\eta^{15}(12z)}{\eta^{11}(2z)},$$

$$f_{16} = \frac{\eta^{14}(4z)\eta^{14}(6z)\eta^{14}(12z)}{\eta^{10}(2z)},$$

$$f_{17} = \frac{\eta^{15}(4z)\eta^{13}(6z)\eta^{19}(12z)}{\eta^{15}(2z)},$$

$$f_{18} = \frac{\eta^{9}(4z)\eta^{19}(6z)\eta^{13}(12z)}{\eta^{9}(2z)},$$

$$f_{19} = \frac{\eta^{18}(4z)\eta^{10}(6z)\eta^{10}(12z)}{\eta^{6}(2z)},$$

$$f_{20} = \frac{\eta^{20}(2z)\eta^{8}(4z)\eta^{20}(6z)}{\eta^{16}(12z)},$$

$$f_{21} = \frac{\eta^{20}(2z)\eta^{20}(4z)}{\eta^{4}(6z)\eta^{4}(12z)},$$

$$f_{22} = \frac{\eta^{20}(4z)\eta^{20}(12z)}{\eta^{4}(2z)\eta^{4}(6z)},$$

$$f_{23} = \frac{\eta^{20}(4z)\eta^{20}(6z)}{\eta^{4}(2z)\eta^{4}(12z)},$$

$$f_{24} = \frac{\eta^{15}(4z)\eta(6z)\eta^{19}(12z)}{\eta^{3}(2z)},$$

$$f_{25} = \frac{\eta^{20}(2z)\eta^{20}(4z)}{\eta^{4}(6z)\eta^{4}(12z)},$$

$$f_{26} = \frac{\eta^{20}(2z)\eta^{20}(6z)\eta^{8}(12z)}{\eta^{4}(4z)},$$

$$f_{27} = \frac{\eta^{19}(2z)\eta^{13}(4z)\eta^{15}(6z)}{\eta^{15}(12z)}.$$

Now we can state our main Theorem:

Theorem 1 Let $b_1, b_2, \cdots, b_5$ be non-negative integers satisfying

$$b_1 + b_2 + \cdots + b_5 \leq 32. \tag{5}$$

Define the integers $a_1, a_2, a_3, a_4, a_6, a_{12}$ by

$$a_1 := -b_1 + 2b_2 - 2b_3 - 4b_4 - b_5 + 32, \tag{6}$$

$$a_2 := 3b_1 + b_2 + 3b_3 + 10b_4 + b_5 - 80, \tag{7}$$

$$a_3 := 3b_1 + 2b_2 + 6b_3 + 4b_4 + 3b_5 - 96, \tag{8}$$

$$a_4 := -2b_1 - b_2 - b_3 - 4b_4 + 2b_5 + 32, \tag{9}$$

$$a_6 := -9b_1 - 7b_2 - 9b_3 - 10b_4 - 7b_5 + 240, \tag{10}$$

$$a_{12} := 6b_1 + 3b_2 + 3b_3 + 4b_4 + 2b_5 - 96. \tag{11}$$

They are functions of q by (3). Now define integers

$k_0, k_1, k_2, k_3, k_4, k_5, k_6, k_7, k_8, k_9, k_{10}, k_{11}, k_{12}, k_{13}, k_{14}, k_{15}, k_{16}, k_{17}, k_{18}, k_{19}, k_{20}, k_{21}, k_{22}, k_{23}, k_{24}, k_{25}, k_{26}, k_{27}, k_{28}, k_{29}, k_{30}, k_{31}$ and k_{32}

by

$$\frac{1}{2^{b_1+b_5}} x^{b_1} (1-x)^{b_2} (1+x)^{b_3} (1+2x)^{b_4} (2+x)^{b_5} \tag{12}$$

$$= k_0 + k_1x + k_2x^2 + k_3x + k_4x^4 + k_5x^5 + k_6x^6 + k_7x^7 + k_8x^8$$

$$+ k_9x^9 + k_{10}x^{10} + k_{11}x^{11} + k_{12}x^{12} + k_{13}x^{13} + k_{14}x^{14} + k_{15}x^{15} \tag{13}$$

$$+k_{16}x^{16}+k_{17}x^{17}+k_{18}x^{18}+k_{19}x^{19}+k_{20}x^{20} \tag{14}$$

$$+k_{21}x^{21}+k_{22}x^{22}+k_{23}x^{23}+k_{24}x^{24}+k_{25}x^{25}+k_{26}x^{26}+k_{27}x^{27} \tag{15}$$

$$+k_{28}x^{28}+k_{29}x^{29}+k_{30}x^{30}+k_{31}x^{31}+k_{32}x^{32}. \tag{16}$$

Define the rational numbers

$c_1, c_2, c_3, c_4, c_6, c_{12}, r_1, r_2, \cdots, r_{26}$

and r_{27} as in Table 1. Here

$\{f_1, \cdots, f_{27}\} \setminus \{f_7, f_9, f_{17}, f_{20}, f_{25}, f_{26}, f_{27}\} \in S_{16}(\Gamma_0(12))$,

$f_7, f_9, f_{17}, f_{20}, f_{25}, f_{26}, f_{27} \in M_{16}(\Gamma_0(12)) \setminus S_{16}(\Gamma_0(12))$ and

$$\eta^{a_1}(z)\eta^{a_2}(2z)\eta^{a_3}(3z)\eta^{a_4}(4z)\eta^{a_6}(6z)\eta^{a_{12}}(12z) = \delta(b_1) + \sum_{n=1}^{\infty} c(n)q^n,$$

where for $n \in \mathbb{N}$,

$$c(n) = -c_1\sigma_{15}(n) - c_2\sigma_{15}\left(\frac{n}{2}\right) - c_3\sigma_{15}\left(\frac{n}{3}\right) - c_4\sigma_{15}\left(\frac{n}{4}\right) - c_6\sigma_{15}\left(\frac{n}{6}\right)$$
$$-c_{12}\sigma_{15}\left(\frac{n}{12}\right) + r_1 f_1(n) + \cdots + r_{27} f_{27}(n).$$

In particular,

$$c(2n) = -c_1\sigma_{15}(2n) - c_2\sigma_{15}(n) - c_4\sigma_{15}\left(\frac{n}{2}\right) - (16385c_3 + c_6)\sigma_{15}\left(\frac{n}{3}\right)$$
$$-(c_{12} - 16384c_3)\sigma_{15}\left(\frac{n}{6}\right) + r_1 f_1(2n) + \cdots + r_{14} f_{14}(2n),$$

$$c(2n-1) = -c_1\sigma_{15}(2n-1) - c_3\sigma_{15}\left(\frac{2n-1}{3}\right) + r_{15} f_{15}(2n-1) + \cdots + r_{27} f_{27}(2n-1),$$

for $n \in \mathbb{N}$.

Table 1. Coefficients of eisenstein series and some eta quotients.

$$c_1 := \frac{72583}{980659548}k_1 - \frac{120832}{245164887}k_0 - \frac{2735}{326886516}k_2 + \frac{277}{490329774}k_3$$
$$-\frac{7}{490329774}k_4 + \frac{1}{36320724}k_5 - \frac{13}{245164887}k_6 + \frac{25}{245164887}k_7$$
$$-\frac{16}{81721629}k_8 + \frac{92}{245164887}k_9 - \frac{16}{22287717}k_{10} + \frac{112}{81721629}k_{11}$$
$$-\frac{640}{245164887}k_{12} + \frac{4352}{245164887}k_{13} - \frac{256}{27240543}k_{14} - \frac{1216}{245164887}k_{15}$$
$$-\frac{8192}{245164887}k_{16} + \frac{5120}{81721629}k_{17} - \frac{28672}{245164887}k_{18} + \frac{16384}{22287717}k_{21}$$
$$-\frac{32768}{81721629}k_{20} - \frac{327680}{245164887}k_{22} + \frac{65536}{27240543}k_{23}$$
$$-\frac{1048576}{245164887}k_{24} + \frac{53248}{245164887}k_{19} + \frac{1835008}{245164887}k_{25}$$
$$-\frac{1048576}{81721629}k_{26} + \frac{5242880}{245164887}k_{27} - \frac{8388608}{245164887}k_{28}$$
$$+\frac{4194304}{81721629}k_{29} - \frac{16777216}{245164887}k_{30} + \frac{16777216}{245164887}k_{31}.$$

$$c_2 := \frac{3473453008873267483251418595008477404924 8}{2142599748818134613113093495072137244191}k_0 - \frac{6295680023718057106647380779677556434730 1}{2571119698581761535735712194086564693029 2}k_1$$
$$+\frac{5605011839556709895971372201606412625 55}{1977784383524431950565932456989665148484}k_2 - \frac{9796420416875499640200489376913398994 9}{4285199497636269226226186990144274488382}k_3$$
$$+\frac{4089707298247402135030526553854251303 7}{1285559849290880767867856097043282346514 6}k_4 - \frac{7791098422160545789710056490553315213 7}{2571119698581761535735712194086564693029 2}k_5$$
$$+\frac{768300741536039363585678834117511034 1}{2142599748818134613113093495072137244191}k_6 - \frac{3206048702036518736069793680488872613 1}{6427799246454403839339280485216411732573}k_7$$
$$+\frac{5062860250020366959225066793546093774 4}{6427799246454403839339280485216411732573}k_8 - \frac{2908526576269334840076990458632147903 6}{2142599748818134613113093495072137244191}k_9$$
$$+\frac{1582503671201287398818468177080744469 92}{6427799246454403839339280485216411732573}k_{10} - \frac{2265691471259882403627662818634073886 4}{4944460958811079876414831142474162871 21}k_{11}$$
$$+\frac{1848487409392476880205547871423990396 16}{2142599748818134613113093495072137244191}k_{12} - \frac{1048239944217959494935262133888301514304}{6427799246454403839339280485216411732573}k_{13}$$

$$+\frac{821035701822957306047778558879567946956 8}{2142599748818134613113093495072137244191}k_{18}-\frac{4574426565126584105107847176407927953408 0}{6427799246454403839339280485216411732573}k_{19}$$

$$+\frac{8445277301233991338040717668231559829913 6}{6427799246454403839339280485216411732573}k_{20}-\frac{5161094884424399679185706551242004763443 2}{2142599748818134613113093495072137244191}k_{21}$$

$$+\frac{2165522546500283181725026991592673607680 0}{494444609588110798764148311424741628 7121}k_{22}-\frac{5067367908741314346570086272492385972387 84}{6427799246454403839339280485216411732573}k_{23}$$

$$+\frac{3002902339887017496979458316335107664773 12}{2142599748818134613113093495072137244191}k_{24}-\frac{1576529708086760827933004700454891591958 528}{6427799246454403839339280485216411732573}k_{25}$$

$$+\frac{2702628877067406586383696467255320708907 008}{6427799246454403839339280485216411732573}k_{26}-\frac{1501462766976347698189587052601574136348 672}{2142599748818134613113093495072137244191}k_{27}$$

$$+\frac{7207028070602459124588069514635005543317 504}{6427799246454403839339280485216411732573}k_{28}-\frac{1081054817007084778924015542964682398485 7088}{6427799246454403839339280485216411732573}k_{29}$$

$$+\frac{4804689379205214665850988696123019241979 904}{2142599748818134613113093495072137244191}k_{30}-\frac{1441406490665108551175926591520290394813 2352}{6427799246454403839339280485216411732573}k_{31}$$

$$-\frac{1261434042760852480251163870389862 4}{1892198777290080612110474090437566009}k_{32},$$

$$
\begin{aligned}
c_3 :=& \frac{32}{2187}k_0 - \frac{62}{2187}k_1 + \frac{40}{729}k_2 - \frac{232}{2187}k_3 + \frac{448}{2187}k_4 \\
&- \frac{32}{81}k_5 + \frac{1664}{2187}k_6 - \frac{3200}{2187}k_7 + \frac{2048}{729}k_8 - \frac{11776}{2187}k_9 \\
&+ \frac{22528}{2187}k_{10} - \frac{14336}{729}k_{11} + \frac{81920}{2187}k_{12} - \frac{155648}{2187}k_{13} \\
&+ \frac{32768}{243}k_{14} - \frac{557056}{2187}k_{15} + \frac{1048576}{2187}k_{16} - \frac{655360}{729}k_{17} \\
&+ \frac{4194304}{729}k_{20} - \frac{23068672}{2187}k_{21} + \frac{41943040}{2187}k_{22} \\
&- \frac{8388608}{243}k_{23} + \frac{134217728}{2187}k_{24} + \frac{3670016}{2187}k_{18} \\
&- \frac{6815744}{2187}k_{19} + \frac{2147483648}{2187}k_{30} + \frac{234881024}{2187}k_{25} \\
&+ \frac{134217728}{729}k_{26} - \frac{671088640}{2187}k_{27} + \frac{1073741824}{2187}k_{28} \\
&- \frac{536870912}{729}k_{29} - \frac{2147483648}{2187}k_{31},
\end{aligned}
$$

$$
\begin{aligned}
c_4 :=& -\frac{34386316699576879433179919982198369120419 84}{2121173751329953266981962560121415871749 09}k_0 + \frac{51939137041571795821951300570913561010176 0}{2121173751329953266981962560121415871749 09}k_1 \\
&- \frac{1541655078827786053038522300113680482304}{5438907054692187864056314256721579158331}k_2 + \frac{4861344345762952626180655745881508651008}{21211737513299532669819625601214158717490 9}k_3 \\
&- \frac{6866622601949659671743441855700099727 36}{21211737513299532669819625601214158717490 9}k_4 + \frac{2422350558926090350565745702361124864 0}{7856199078999826914748009481931169895367}k_5 \\
&- \frac{77091290577603993304304035276554351411 2}{21211737513299532669819625601214158717490 9}k_6 + \frac{10666740408738841951450725154313900195 84}{21211737513299532669819625601214158717490 9}k_7 \\
&- \frac{5589094248144163468551534772627570688 00}{7070579171099844223273208533738052905830 3}k_8 + \frac{28809434473100056572719008241993374105 60}{21211737513299532669819625601214158717490 9}k_9 \\
&- \frac{4742121168681648179107235640623311093 76}{1928339773936321151801784145564923519771 9}k_{10} + \frac{2487599390952252563307370096133535825 92}{5438907054692187864056314256721579158331}k_{11} \\
&- \frac{18261794573801688749784466114240638353408}{21211737513299532669819625601214158717490 9}k_{12} + \frac{34521130653913088315205006307078765740032}{21211737513299532669819625601214158717490 9}k_{13}
\end{aligned}
$$

$$-\frac{4742121168681648179107235640623311093 76}{192833977393632115180178414556492351977 19}k_{10}+\frac{2487599390952252563307370096133535825 92}{543890705469218786405631425672157915833 1}k_{11}$$

$$-\frac{182617945738016887497844661142406383534 08}{212117375132995326698196256012141587174909}k_{12}+\frac{345211306539130883152050063070787657400 32}{212117375132995326698196256012141587174909}k_{13}$$

$$-\frac{725297297036827298991850520539485136486 4}{235685972369994807442440284457935096861 01}k_{14}+\frac{123219038849645502527071198852093878009 856}{212117375132995326698196256012141587174909}k_{15}$$

$$-\frac{231919667081212951123368487557465697681408}{212117375132995326698196256012141587174909}k_{16}+\frac{14496722265561367397984011687291015246643 2}{70705791710998442232732085337380529058303}k_{17}$$

$$-\frac{811966565987547701423795536752529381916672}{212117375132995326698196256012141587174909}k_{18}+\frac{150822201543843266409159500971363174665420 8}{212117375132995326698196256012141587174909}k_{19}$$

$$-\frac{928294070025214438646146335719133007839232}{70705791710998442232732085337380529058303}k_{20}+\frac{464214257594362122128001913465962465394688}{192833977393632115180178414556492351977 19}k_{21}$$

$$-\frac{714261708179731033246971232677038694989824}{16316721164076563592168942770164737474993}k_{22}+\frac{1857262903145106340717278218703097268338688}{235685972369994807442440284457935096861 01}k_{23}$$

$$-\frac{29718557905407543679170452853863882287480832}{212117375132995326698196256012141587174909}k_{24}+\frac{5201074375383959516944740864358781274364313 6}{212117375132995326698196256012141587174909}k_{25}$$

$$-\frac{29721880606491798348403071893001785716637696}{70705791710998442232732085337380529058303}k_{26}+\frac{1486149076595838002782198309592236301463715 84}{212117375132995326698196256012141587174909}k_{27}$$

$$-\frac{23779005342560474904583209077318236522754867 2}{212117375132995326698196256012141587174909}k_{28}+\frac{118896676835312618271459427865760036870225920}{70705791710998442232732085337380529058303}k_{29}$$

$$-\frac{47558337855697422303129880603573220151040409 6}{212117375132995326698196256012141587174909}k_{30}+\frac{475544908027362753251165705288941682960629760}{212117375132995326698196256012141587174909}k_{31}$$

$$+\frac{6966286421316436981768488433064345 6}{700814361959289115596471885347246 67}k_{32}.$$

$$
\begin{aligned}
c_6 := & -\frac{9031408234257645392661757427811295633328}{1892198777290080612110474090437566009}k_0 + \frac{1753599251608810820042292756595554516256}{1892198777290080612110474090437566009}k_1 \\
& -\frac{261382625829009040606391162572202246708}{145553752099236970162344160802889693}k_2 + \frac{6573369312471681371367564937022266290064}{1892198777290080612110474090437566009}k_3 \\
& -\frac{12697362031501480563543217533293921843408}{1892198777290080612110474090437566009}k_4 + \frac{2449172368082673174245281050849282977 8096}{1892198777290080612110474090437566009}k_5 \\
& -\frac{4717319030318709689007625501684918013 5904}{1892198777290080612110474090437566009}k_6 + \frac{9072160715646935467176767808675679540 9024}{1892198777290080612110474090437566009}k_7 \\
& -\frac{174189406811691020942065915513466611942400}{1892198777290080612110474090437566009}k_8 + \frac{33386693721644068072322005810086147037 3888}{1892198777290080612110474090437566009}k_9 \\
& -\frac{638705859516034234832558641988331987268352}{1892198777290080612110474090437566009}k_{10} + \frac{93796263570379824927121966639465976592640}{145553752099236970162344160802889693}k_{11} \\
& -\frac{2322578004133601510600390747954985709790720}{1892198777290080612110474090437566009}k_{12} + \frac{4412902046734060157362588620252943713253888}{1892198777290080612110474090437566009}k_{13} \\
& -\frac{8361291905582645393038766948930775507203072}{1892198777290080612110474090437566009}k_{14} \\
& +\frac{15793555169896537058264823055463258597435392}{1892198777290080612110474090437566009}k_{15} \\
& -\frac{29729048791079112416901620246445683858493440}{1892198777290080612110474090437566009}k_{16} \\
& +\frac{55741970217875135338106642659799456846036992}{1892198777290080612110474090437566009}k_{17} \\
& -\frac{104051681439650108926296625737844180389232640}{1892198777290080612110474090437566009}k_{18} \\
& +\frac{193238840618886918635723645699723731688488960}{1892198777290080612110474090437566009}k_{19} \\
& -\frac{356748632448054978156317427837374824256110592}{1892198777290080612110474090437566009}k_{20} \\
& +\frac{654039163047100217481579906467314595064709120}{1892198777290080612110474090437566009}k_{21} \\
& -\frac{914740090866098785191698590489748308329758 72}{145553752099236970162344160802889693}k_{22}
\end{aligned}
$$

$$
\begin{aligned}
&+\frac{2140491816044378587791643159133715725272612864}{189219877729008061211047409043756600 9}k_{23}\\
&-\frac{3805318787402182990247254077934953884303753216}{1892198777290080612110474090437566009}k_{24}\\
&+\frac{6659307881158913920815458330065825499748892672}{1892198777290080612110474090437566009}k_{25}\\
&-\frac{11415956370753930456327232934139762697522380800}{1892198777290080612110474090437566009}k_{26}\\
&+\frac{19026593954106378493996218523841248653172670464}{1892198777290080612110474090437566009}k_{27}\\
&-\frac{30442550329135403612241707744616567638358228992}{1892198777290080612110474090437566009}k_{28}\\
&+\frac{45663825495841001129663585013837084183763222528}{1892198777290080612110474090437566009}k_{29}\\
&-\frac{60885100662546560092483434885863179799076012032}{1892198777290080612110474090437566009}k_{30}\\
&+\frac{60885100662544944606303920653576106827577294848}{1892198777290080612110474090437566009}k_{31}\\
&+\frac{856104443466901459820484315334259 5072}{1892198777290080612110474090437566009}k_{32},
\end{aligned}
$$

$$
\begin{aligned}
c_{12} :=\ & \frac{32528099158590735197773034014311108157440}{18921987772900806121104740904375660009} k_0 - \frac{812779358899696260571931072917630803968}{6307329257633602040368246968125220003} k_1 \\
& + \frac{28882881911778850460194128433207717880}{1455537520992369701623441608028896930} k_2 - \frac{68617697184175422326599708328487391311392}{18921987772900806121104740904375660009} k_3 \\
& + \frac{43196402138715835822525490718870977088544}{6307329257633602040368246968125220003} k_4 - \frac{247310108747508720361037729965044980940800}{18921987772900806121104740904375660009} k_5 \\
& + \frac{47390045096215471516305801919735885168640}{18921987772900806121104740904375660009} k_6 - \frac{10101702184638558192238504007106640936960}{2102443085877867346789415656041740001} k_7 \\
& + \frac{17435874331565098372468343509911778990489 6}{18921987772900806121104740904375660009} k_8 - \frac{33400956955255000329643913757794908576153 6}{18921987772900806121104740904375660009} k_9 \\
& + \frac{21293911919917370254912386008413467285913 6}{6307329257633602040368246968125220003} k_{10} - \frac{9380179796392406613183535986382161910169 6}{1455537520992369701623441608028896930} k_{11} \\
& + \frac{2322594452836691591464083034791767560945664}{18921987772900806121104740904375660009} k_{12} - \frac{1470944291358380225706902495347555354083328}{6307329257633602040368246968125220003} k_{13} \\
& + \frac{836108040955255742224478721942541193877913 6}{18921987772900806121104740904375660009} k_{14} \\
& - \frac{1579309503424882692197184240537593493225472 0}{18921987772900806121104740904375660009} k_{15} \\
& + \frac{1101042279957564750138573655175586727854080}{700814361959289115596471885347246670} k_{16} \\
& - \frac{5574024732620284551658189112569523533296435 2}{18921987772900806121104740904375660009} k_{17} \\
& + \frac{10404846246485658774427265980588551254717235 2}{18921987772900806121104740904375660009} k_{18} \\
& - \frac{6441095937256854455858822185360093046944563 2}{6307329257633602040368246968125220003} k_{19} \\
& + \frac{3567376583420271341208028110297318719307120 64}{18921987772900806121104740904375660009} k_{20} \\
& - \frac{6540190947953114128769845932106709082058915 84}{18921987772900806121104740904375660009} k_{21} \\
& + \frac{3049040250970050315674592351901001198482227 2}{\text{[illegible]}} k_{22}
\end{aligned}
$$

$$
\begin{aligned}
&-\frac{2140426342563992719357502322247849299974094848}{1892198777290080612110474090437566009}k_{23}\\
&+\frac{3805202487140833038817416109251596329128820736}{1892198777290080612110474090437566009}k_{24}\\
&-\frac{7399004960967512164484573732298530463011962288}{210244308587786734678941565604174001}k_{25}\\
&+\frac{1141560777564696109444959746198858984405480243 2}{1892198777290080612110474090437566009}k_{26}\\
&-\frac{1902601308599373280639282520432034392358289408 0}{1892198777290080612110474090437566009}k_{27}\\
&+\frac{1014720702081141212052358662438819554306424832 0}{630732925763360204036824696812522003}k_{28}\\
&-\frac{4566243170496131866568152887773305430588594585 6}{1892198777290080612110474090437566009}k_{29}\\
&+\frac{6088324234744766386881079389379174151357058252 8}{1892198777290080612110474090437566009}k_{30}\\
&-\frac{2029441410118825862656021109535873142807959961 6}{630732925763360204036824696812522003}k_{31}\\
&-\frac{70829276182219349598717929801032663040 0}{1892198777290080612110474090437566009}k_{32}.
\end{aligned}
$$

$$r_1 := \frac{91310053715411127712}{63561267}k_0 - \frac{7641452474572136509}{36320724}k_1 + \frac{1237277739610086155}{84748356}k_2 + \frac{301998497851066193}{18160362}k_3 - \frac{4445149892148477245}{127122534}k_4 + \frac{634162625358701723}{9416484}k_5 - \frac{8243024078809732703}{63561267}k_6 + \frac{15852441259742139299}{63561267}k_7 - \frac{10145856761562783152}{21187089}k_8 + \frac{58339579743384278452}{63561267}k_9 - \frac{10146087995299300784}{5778297}k_{10} + \frac{71022855408522794960}{21187089}k_{11} - \frac{405845531481576392576}{63561267}k_{12} + \frac{771107108760819174464}{63561267}k_{13} - \frac{162338403816018689792}{7062363}k_{14} + \frac{2759753469532240738048}{63561267}k_{15} - \frac{5194830713883325932544}{63561267}k_{16} + \frac{3246769440142703828992}{21187089}k_{17} - \frac{18181909703056191509504}{63561267}k_{18} - \frac{20779326348351079361536}{21187089}k_{20} + \frac{33766404705203191651328}{63561267}k_{19} + \frac{10389663292872757535744}{5778297}k_{21} - \frac{207793267358083979723776}{63561267}k_{22} + \frac{41558653661276242758656}{7062363}k_{23} - \frac{664938460496776958624768}{63561267}k_{24} + \frac{1163642307985908317785088}{63561267}k_{25} - \frac{664938462468936095771648}{21187089}k_{26} + \frac{3324692314743380900555776}{63561267}k_{27} - \frac{5319507705979530190849024}{63561267}k_{28} + \frac{2659753853705085386064896}{21187089}k_{29} - \frac{10639015416250982125540352}{63561267}k_{30} + \frac{10639015416250982125540352}{63561267}k_{31},$$

$$r_2 := \frac{12586505097374334976}{190683801}k_0 - \frac{3733982941123692544}{27240543}k_1 + \frac{16889620877821442048}{63561267}k_2 - \frac{13972654485209821184}{27240543}k_3 + \frac{188725357515715014656}{190683801}k_4 - \frac{13476065128386856960}{7062363}k_5$$

$$
\begin{aligned}
&+\frac{7006656916474899974272}{190683801}k_{6}-\frac{1347366543765495291904}{190683801}k_{7}+\frac{862298199446149267456}{63561267}k_{8}\\
&-\frac{4958180642825882009600}{190683801}k_{9}+\frac{862290353928014528512}{17334891}k_{10}-\frac{603602908792758717 6448}{63561267}k_{11}\\
&+\frac{3449159474803233154 6624}{190683801}k_{12}-\frac{6553403986291169689 6000}{190683801}k_{13}+\frac{1379664221139152876 3392}{21187089}k_{14}\\
&-\frac{2345429491756749852 50816}{190683801}k_{15}+\frac{4414926549413938127 17568}{190683801}k_{16}-\frac{2759329294002607625 33888}{63561267}k_{17}\\
&+\frac{1765970941189026648 424448}{63561267}k_{20}+\frac{1545224483122349019 234304}{190683801}k_{18}-\frac{2869702711575934430 347264}{190683801}k_{19}\\
&-\frac{8829854846217094310 13376}{17334891}k_{21}+\frac{1765970987954122699 9906304}{190683801}k_{22}\\
&-\frac{3531942000722552879 054848}{21187089}k_{23}+\frac{5651107227363828467 8660096}{190683801}k_{24}\\
&-\frac{9889437678048979952 3368960}{190683801}k_{25}+\frac{5651107255892379423 1066624}{63561267}k_{26}\\
&-\frac{2825553631623681124 61348864}{190683801}k_{27}+\frac{4520885814371650671 01011968}{190683801}k_{28}\\
&-\frac{2260442908337207488 70729728}{63561267}k_{29}+\frac{9041771635651594261 23366400}{190683801}k_{30},
\end{aligned}
$$

$$
\begin{aligned}
r_3 := {}& \frac{15566728830133141504}{63561267}k_0 + \frac{3185223803802372608}{9080181}k_1 - \frac{17041426163914032640}{21187089}k_2 \\
&+ \frac{14256037267597450240}{9080181}k_3 - \frac{19281699648055816 4992}{63561267}k_4 + \frac{13774493971286120960}{2354121}k_5 \\
&- \frac{71631658282411637 3504}{63561267}k_6 + \frac{137756793945881609 2160}{63561267}k_7 - \frac{88165370922902857 3184}{21187089}k_8 \\
&+ \frac{506953570569091556 1472}{63561267}k_9 - \frac{88166062339483398 9632}{5778297}k_{10} + \frac{617163237291059403 1616}{21187089}k_{11} \\
&- \frac{352664958653925431 70560}{63561267}k_{12} + \frac{670063703551437547 76576}{63561267}k_{13} - \frac{141066079944992608 74752}{7062363}k_{14} \\
&+ \frac{239812376836928988 774400}{63561267}k_{15} - \frac{451411584006449932 926976}{63561267}k_{16} + \frac{282132261439511343 923200}{21187089}k_{17} \\
&- \frac{157994074480987834 5089024}{63561267}k_{18} - \frac{180564666902670641 4764032}{21187089}k_{20} + \frac{293417576968949356 2982400}{63561267}k_{19} \\
&+ \frac{902823348363955189 317632}{5778297}k_{21} - \frac{180564671513558930 15511040}{63561267}k_{22} + \frac{361129345465208362 2379520}{7062363}k_{23} \\
&- \frac{577806955320499340 81982464}{63561267}k_{24} + \frac{101116217478105169 134092288}{63561267}k_{25} \\
&- \frac{577806958132865278 73712128}{21187089}k_{26} + \frac{288903479430757697 593016320}{63561267}k_{27} \\
&- \frac{462245567464244031 099633664}{63561267}k_{28} + \frac{231122783846772861 380329472}{21187089}k_{29} \\
&- \frac{924491135616393137 182343168}{63561267}k_{30} + \frac{924491135616393137 182343168}{63561267}k_{31},
\end{aligned}
$$

$$
\begin{aligned}
r_4 := {}& -\frac{13192965750455434 5952}{190683801}k_0 - \frac{77985833771912592 11}{36320724}k_1 + \frac{49721424993841598 0821}{762735204}k_2 \\
&- \frac{70199075741890437 883}{54481086}k_3 + \frac{95018720090660602 5247}{381367602}k_4 - \frac{40734886214822069 5163}{84748356}k_5 \\
&+ \frac{58847442011489840 7047}{63561267}k_6 - \frac{33952498202898637 76129}{190683801}k_7 + \frac{65190506339564231 21840}{190683801}k_8
\end{aligned}
$$

$$
\begin{aligned}
&-\frac{12494990286877563773116}{190683801}k_{9}+\frac{241450319017769846288}{1926099}k_{10}-\frac{45634217333431409584592}{190683801}k_{11}\\
&+\frac{86922417480145120334464}{190683801}k_{12}-\frac{165152690405816521254592}{190683801}k_{13}+\frac{104306996560416660578048}{63561267}k_{14}\\
&-\frac{197024365259424744775424}{63561267}k_{15}+\frac{111260831590885708452966 4}{190683801}k_{16}-\frac{208614075649460449678950 4}{190683801}k_{17}\\
&+\frac{389412961364847465856102 4}{190683801}k_{18}-\frac{241065174839315987382169 6}{63561267}k_{19}\\
&+\frac{13351302296637597451316224}{190683801}k_{20}-\frac{2225217083402369781173248}{17334891}k_{21}\\
&+\frac{44504342121021691370875904}{190683801}k_{22}-\frac{26702605453785736254491648}{63561267}k_{23}\\
&+\frac{15823766266379980287331328}{21187089}k_{24}-\frac{249224319446065870083802112}{190683801}k_{25}\\
&+\frac{427241691334835494525561856}{190683801}k_{26}-\frac{712069486501952645203108864}{190683801}k_{27}\\
&+\frac{379770393128547090469430272}{63561267}k_{28}-\frac{1708966769985901373141107712}{190683801}k_{29}\\
&+\frac{2278622360586161474873924608}{190683801}k_{30}-\frac{2278622360586161474873924608}{190683801}k_{31},
\end{aligned}
$$

$$
\begin{aligned}
r_5 := &-\frac{4078493266508992}{3026727}k_0 + \frac{1757093202559117}{672606}k_1 - \frac{30608970079163071}{6053454}k_2 + \frac{29588812154023063}{3026727}k_3 \\
&-\frac{5713711904020917}{3026727}k_4 + \frac{24487373790016241}{672606}k_5 - \frac{23580465846292238}{336303}k_6 + \frac{408123936542458630}{3026727}k_7 \\
&-\frac{783598788593292832}{3026727}k_8 + \frac{1501899078618026600}{3026727}k_9 - \frac{87066684321864992}{91719}k_{10} \\
&+\frac{5485205009982032608}{3026727}k_{11} - \frac{10448015967673411328}{3026727}k_{12} + \frac{19851240835474624640}{3026727}k_{13} \\
&-\frac{12537631452645220864}{1008909}k_{14} + \frac{23682201810197115392}{1008909}k_{15} - \frac{133734829754049019904}{3026727}k_{16} \\
&+\frac{250752873123348457472}{3026727}k_{17} - \frac{468072133707326267392}{3026727}k_{18} + \frac{289758992548733034496}{1008909}k_{19} \\
&-\frac{1604819271844457676800}{3026727}k_{20} + \frac{267469910195425869824}{275157}k_{21} \\
&-\frac{5349398704186073153536}{3026727}k_{22} + \frac{3209639457729863090176}{1008909}k_{23} \\
&-\frac{5706026026397239083008}{1008909}k_{24} + \frac{29956637937551790309376}{3026727}k_{25} \\
&-\frac{51354238139489320960000}{3026727}k_{26} + \frac{85590398952084000997376}{3026727}k_{27} \\
&-\frac{15216071182490613579776}{336303}k_{28} + \frac{205416963210483884621824}{3026727}k_{29} \\
&-\frac{273889285778552247025664}{3026727}k_{30} + \frac{273889285778552247025664}{3026727}k_{31},
\end{aligned}
$$

$$
\begin{aligned}
r_6 := &\frac{1785053112534131 4656}{190683801}k_0 - \frac{15602348027846377731}{108962172}k_1 + \frac{565694547181365397}{254245068}k_2 \\
&-\frac{757706523423 62225}{54481086}k_3 + \frac{954011446348337117}{381367602}k_4 - \frac{136502477859918587}{28249452}k_5 \\
&+\frac{1775145743398433855}{190683801}k_6 - \frac{3414079397677526531}{190683801}k_7 + \frac{2185090542419623856}{63561267}k_8
\end{aligned}
$$

$$
\begin{aligned}
&-\frac{12564463935821044276}{190683801}k_9+\frac{2185139107269117872}{17334891}k_{10}-\frac{1529602169529610798 4}{63561267}k_{11} \\
&+\frac{8740596824103440883 2}{190683801}k_{12}-\frac{16607146221443229497 6}{190683801}k_{13}+\frac{3496242659144669465 6}{21187089}k_{14} \\
&-\frac{5943613785584163827 20}{190683801}k_{15}+\frac{11187980266193400217 60}{190683801}k_{16}-\frac{6992488181113986979 84}{63561267}k_{17} \\
&+\frac{3915793558221726961664}{190683801}k_{18}-\frac{447519284258976396083 2}{63561267}k_{20} \\
&-\frac{2237596446112370425856}{17334891}k_{21}+\frac{7272188241043282374656}{190683801}k_{19} \\
&+\frac{4475192923490718937907 2}{190683801}k_{22}-\frac{8950385886365732618240}{21187089}k_{23} \\
&+\frac{14320617457857088323584 0}{190683801}k_{24}-\frac{2506108059494356119879 68}{190683801}k_{25} \\
&+\frac{14320617498530720340377 6}{63561267}k_{26}-\frac{7160308754194099850936 32}{190683801}k_{27} \\
&+\frac{11456494011612818438225 92}{190683801}k_{28}+\frac{5728247007271790621327 36}{63561267}k_{29} \\
&-\frac{22912988032017925289738 24}{190683801}k_{30}-\frac{22912988032017925289738 24}{190683801}k_{31},
\end{aligned}
$$

$$r_7 := \frac{1829852116193665024 00}{27240543}k_0 - \frac{9114335538551103232}{9080181}k_1 + \frac{3015119186275104512}{27240543}k_2 - \frac{1605771279329520 64}{27240543}k_3 - \frac{2080586500978841 6}{27240543}k_4 + \frac{593806988273408}{3026727}k_5 + \frac{2355269316932608}{9080181}k_6 - \frac{22249981901335552}{27240543}k_7 + \frac{47041546286743552}{27240543}k_8 - \frac{923186815594864 64}{27240543}k_9 + \frac{1794795498225664}{275157}k_{10} - \frac{339752396451659776}{27240543}k_{11} + \frac{647414622429839360}{27240543}k_{12} - \frac{1230220929648558080}{27240543}k_{13} + \frac{777003747655745536}{9080181}k_{14} - \frac{1467684753032151040}{9080181}k_{15} + \frac{8288118568652177408}{27240543}k_{16} - \frac{15540230489066110976}{27240543}k_{17} + \frac{29008434308301783040}{27240543}k_{18} - \frac{17957602863013756928}{9080181}k_{19} + \frac{99457493780231684096}{27240543}k_{20} - \frac{16576249008379461632}{2476413}k_{21} + \frac{331524980412700229632}{27240543}k_{22} - \frac{198914988287781830656}{9080181}k_{23} + \frac{117875548621602881536}{3026727}k_{24} - \frac{1856539890819019767808}{27240543}k_{25} + \frac{3182639812846948777984}{27240543}k_{26} - \frac{5304399688084778319872}{27240543}k_{27} + \frac{2829013166979691249664}{9080181}k_{28} - \frac{16974079001880596381696}{27240543}k_{30} - \frac{16974079001880596381696}{27240543}k_{31} + \frac{12730559251409835065344}{27240543}k_{29},$$

$$r_8 := \frac{1402046646636032}{572051403}k_0 - \frac{404359709107292}{81721629}k_1 + \frac{1838528360224516}{190683801}k_2 - \frac{1526510201939944}{81721629}k_3 + \frac{20651859976599112}{572051403}k_4 - \frac{1475682550603964}{21187089}k_5 + \frac{76747948993198640}{572051403}k_6 - \frac{147602599162168688}{572051403}k_7 + \frac{94468543272975104}{190683801}k_8$$

$$
\begin{aligned}
&-\frac{543201327526722112}{572051403}k_{9}+\frac{94470348795976448}{52004673}k_{10}-\frac{661294196377898240}{190683801}k_{11}\\
&+\frac{377882873329070 7968}{572051403}k_{12}-\frac{7179779137236426752}{572051403}k_{13}+\frac{1511532964276548608}{63561267}k_{14}\\
&-\frac{25696065395082001408}{572051403}k_{15}+\frac{48369069929078318080}{572051403}k_{16}-\frac{30230670898864546816}{190683801}k_{17}\\
&+\frac{19347631187845965 7216}{190683801}k_{20}+\frac{169291764793840182272}{572051403}k_{18}\\
&-\frac{314399000941203786752}{572051403}k_{19}-\frac{96738157089236403200}{52004673}k_{21}\\
&+\frac{1934763156414028899328}{572051403}k_{22}-\frac{386952633138707606528}{63561267}k_{23}\\
&+\frac{6191242149010792604672}{572051403}k_{24}-\frac{10834673781543178923008}{572051403}k_{25}\\
&+\frac{6191242168369447930880}{190683801}k_{26}-\frac{30956210865390042385408}{572051403}k_{27}\\
&+\frac{49529937408072029873152}{572051403}k_{28}-\frac{24764968711051435418624}{190683801}k_{29}\\
&+\frac{99059874858236582638592}{572051403}k_{30}-\frac{99059874858236582638592}{572051403}k_{31}.
\end{aligned}
$$

$$
\begin{aligned}
r_9 := {} & \frac{410440630890496}{245164887}k_0 - \frac{822808296656416}{245164887}k_1 + \frac{532781926617632}{81721629}k_2 \\
& - \frac{3092465057117888}{245164887}k_3 + \frac{5973557948909120}{245164887}k_4 - \frac{426751782276064}{9080181}k_5 \\
& + \frac{22192952299424128}{245164887}k_6 - \frac{42680608321127296}{245164887}k_7 + \frac{27316205232019456}{81721629}k_8 \\
& - \frac{157070021108969984}{245164887}k_9 + \frac{27316692114626560}{22287717}k_{10} - \frac{191217453405644800}{81721629}k_{11} \\
& + \frac{1092672979433414656}{245164887}k_{12} - \frac{2076080468475166720}{245164887}k_{13} + \frac{437069771975139328}{27240543}k_{14} \\
& - \frac{7430187908813594624}{245164887}k_{15} + \frac{13986237835758608384}{245164887}k_{16} - \frac{8741399233129766912}{81721629}k_{17} \\
& + \frac{48951837446132015104}{245164887}k_{18} - \frac{5594495870713386598 4}{81721629}k_{20} \\
& - \frac{279724795057177477 12}{22287717}k_{21} + \frac{90910556978580103168}{245164887}k_{19} \\
& + \frac{559449591757583753216}{245164887}k_{22} - \frac{111889918530040373248}{27240543}k_{23} \\
& + \frac{1790238698042727669760}{245164887}k_{24} - \frac{3132917723081066487808}{245164887}k_{25} \\
& + \frac{1790238699381654839296}{81721629}k_{26} - \frac{8951193498247201366016}{245164887}k_{27} \\
& + \frac{14321909598400556638208}{245164887}k_{28} + \frac{7160954799535010111488}{81721629}k_{29} \\
& + \frac{28643819198809504030720}{245164887}k_{30} - \frac{28643819198809504030720}{245164887}k_{31},
\end{aligned}
$$

$$
\begin{aligned}
r_{10} := {} & \frac{497136946726582528}{190683801}k_0 - \frac{101223662133026 74}{27240543}k_1 + \frac{487378431477950}{63561267}k_2 \\
& + \frac{1798343394456628}{27240543}k_3 - \frac{25298324783503396}{190683801}k_4 + \frac{1806651219262046}{7062363}k_5 \\
& - \frac{93955276958445080}{190683801}k_6 + \frac{180694470620880824}{190683801}k_7 - \frac{115647804424121216}{63561267}k_8
\end{aligned}
$$

$$
\begin{aligned}
&+\frac{6649835110540352 32}{190683801}k_{9}-\frac{115649988633594752}{17334891}k_{10}+\frac{809552148599316608}{63561267}k_{11} \\
&-\frac{4626018445010825216}{190683801}k_{12}+\frac{8789440995378177536}{190683801}k_{13}-\frac{1850409300232641536}{21187089}k_{14} \\
&+\frac{3145696449485264 5888}{190683801}k_{15}-\frac{59213116671659852800}{190683801}k_{16}+\frac{3700820056858516 7872}{63561267}k_{17} \\
&-\frac{20724593229807979 8272}{190683801}k_{18}-\frac{23685250457641868 8000}{63561267}k_{20}+\frac{384885313359433100288}{190683801}k_{19} \\
&+\frac{118426253556555441152}{17334891}k_{21}-\frac{2368525087038476502016}{190683801}k_{22}+\frac{473705019402448833536}{21187089}k_{23} \\
&-\frac{7579280330445604668416}{190683801}k_{24}+\frac{13263740600225661307904}{190683801}k_{25} \\
&-\frac{7579280350845647793152}{63561267}k_{26}+\frac{37896401778812429495296}{190683801}k_{27} \\
&-\frac{60634242870485098267648}{190683801}k_{28}+\frac{30317121442518316565504}{63561267}k_{29} \\
&-\frac{121268485784624801125376}{190683801}k_{30}+\frac{121268485784624801125376}{190683801}k_{31},
\end{aligned}
$$

$$r_{11} := -\frac{7845155552}{245164887}k_0 + \frac{1961246513}{980659548}k_1 + \frac{2735}{326886516}k_2 - \frac{277}{490329774}k_3 + \frac{7}{490329774}k_4 - \frac{1}{36320724}k_5$$
$$+\frac{13}{245164887}k_6 - \frac{25}{245164887}k_7 + \frac{16}{81721629}k_8 - \frac{92}{245164887}k_9 + \frac{16}{22287717}k_{10} - \frac{112}{81721629}k_{11} + \frac{640}{245164887}k_{12}$$
$$-\frac{1216}{245164887}k_{13} + \frac{256}{27240543}k_{14} - \frac{4352}{245164887}k_{15} + \frac{8192}{245164887}k_{16} - \frac{5120}{81721629}k_{17} + \frac{28672}{245164887}k_{18} - \frac{32768}{81721629}k_{20}$$
$$-\frac{16384}{22287717}k_{21} + \frac{327680}{245164887}k_{22} - \frac{65536}{27240543}k_{23} + \frac{53248}{245164887}k_{19} + \frac{1048576}{245164887}k_{24} - \frac{1835008}{245164887}k_{25} + \frac{1048576}{81721629}k_{26}$$
$$-\frac{5242880}{245164887}k_{27} + \frac{8388608}{245164887}k_{28} - \frac{16777216}{245164887}k_{30} - \frac{16777216}{245164887}k_{31} + \frac{4194304}{81721629}k_{29};$$

$$r_{12} := \frac{2317220714287552}{190683801}k_0 + \frac{152166700793161}{54481086}k_1 - \frac{1195931905993439}{127122534}k_2$$
$$+\frac{508464732469555}{27240543}k_3 - \frac{6883955676246967}{190683801}k_4 + \frac{983776951113553}{14124726}k_5$$
$$-\frac{25582341097777850}{190683801}k_6 + \frac{49200272990040146}{190683801}k_7 - \frac{31489134650021408}{63561267}k_8$$
$$+\frac{181064925473328184}{190683801}k_9 - \frac{31489736491021856}{17334891}k_{10} + \frac{220428740372507360}{63561267}k_{11}$$
$$-\frac{1259594386791424256}{190683801}k_{12} + \frac{2393230849565066624}{190683801}k_{13} - \frac{503838245036658176}{21187089}k_{14}$$
$$+\frac{8565251833083414016}{190683801}k_{15} - \frac{16122828865249315840}{190683801}k_{16} + \frac{10076768771844355072}{63561267}k_{17}$$
$$-\frac{56429907709061295104}{190683801}k_{18} + \frac{3224566347419188736 0}{17334891}k_{21} - \frac{64491326181712724992}{63561267}k_{20}$$
$$+\frac{104798403091518792704}{190683801}k_{19} - \frac{644913274360271358976}{190683801}k_{22} + \frac{128982655490688214016}{21187089}k_{23}$$
$$-\frac{206372249411483505766 4}{190683801}k_{24} + \frac{361151437162572530585 6}{190683801}k_{25}$$
$$-\frac{2063722500567720166400}{63561267}k_{26} + \frac{1031861251068620174233 6}{190683801}k_{27}$$
$$-\frac{1650978002491391013990 4}{190683801}k_{28} + \frac{825489001479542856396 8}{63561267}k_{29}$$

$$-\frac{3301956006385866 1243904}{190683801}k_{30}+\frac{33019560063858661243904}{190683801}k_{31},$$

$$r_{13}:=\frac{1713565605536}{3026727}k_0-\frac{4885835317153}{4035636}k_1+\frac{28580433958345}{12106908}k_2-\frac{27640118395825}{6053454}k_3+\frac{53375164694971}{6053454}k_4$$
$$-\frac{22875194940935}{1345212}k_5+\frac{33042120966595}{1008909}k_6-\frac{190628577556325}{3026727}k_7+\frac{366008629043312}{3026727}k_8-\frac{701519765914156}{3026727}k_9$$
$$+\frac{40667991081200}{91719}k_{10}-\frac{2562094151981840}{3026727}k_{11}+\frac{4880198724104320}{3026727}k_{12}-\frac{9272412472391104}{3026727}k_{13}$$
$$+\frac{5856281324390144}{1008909}k_{14}-\frac{3687300576400640}{336303}k_{15}+\frac{62467405675601920}{3026727}k_{16}-\frac{117126728485473280}{3026727}k_{17}$$
$$+\frac{218637154101997568}{3026727}k_{18}-\frac{15038572140081152}{112101}k_{19}+\frac{749616704535101440}{3026727}k_{20}-\frac{124936378636779520}{275157}k_{21}$$
$$+\frac{2498732274592317440}{3026727}k_{22}-\frac{1499241872412311552}{1008909}k_{23}+\frac{2665322785088208896}{1008909}k_{24}-\frac{13992962175311544320}{3026727}k_{25}$$
$$+\frac{23987960234246144000}{3026727}k_{26}-\frac{39979967159169187840}{3026727}k_{27}+\frac{21322662525727145984}{1008909}k_{28}$$
$$+\frac{95952021488282894336}{3026727}k_{29}-\frac{127936055399384350720}{3026727}k_{30}-\frac{127936055399384350720}{3026727}k_{31},$$

$$r_{14} := \frac{8258019327389696}{190683801}k_0 - \frac{2364911193971552}{27240543}k_1 + \frac{10718724919702432}{63561267}k_2 - \frac{8886997089409600}{27240543}k_3 + \frac{12015452534199424}{190683801}k_4$$
$$- \frac{8583279015924320}{7062363}k_5 + \frac{446349030653909888}{190683801}k_6 - \frac{858379450288459136}{190683801}k_7 + \frac{549367570134431744}{63561267}k_8 - \frac{3158876750178781696}{190683801}k_9$$
$$+ \frac{549371059459782656}{17334891}k_{10} - \frac{3845602001029695488}{63561267}k_{11} + \frac{21974883783699685376}{190683801}k_{12} - \frac{41752296617398059008}{190683801}k_{13}$$
$$+ \frac{8789959449390325760}{21187089}k_{14} - \frac{149429334740324589568}{190683801}k_{15} + \frac{281278776244639277056}{190683801}k_{16} - \frac{175799246338520276992}{63561267}k_{17}$$
$$+ \frac{1125115266765590093824}{63561267}k_{20} + \frac{984475818726279618560}{190683801}k_{18} - \frac{1828312280237141762048}{190683801}k_{19} - \frac{562557638768933888000}{17334891}k_{21}$$
$$+ \frac{11251152841939587506176}{190683801}k_{22} - \frac{2250230576600004141056}{21187089}k_{23} + \frac{36003689306605160013824}{190683801}k_{24} - \frac{63006456373979816468480}{190683801}k_{25}$$
$$+ \frac{36003689387387099242496}{63561267}k_{26} - \frac{180018447031999325249536}{190683801}k_{27} + \frac{288029515344343620493312}{190683801}k_{28}$$
$$- \frac{144014757699731394486272}{63561267}k_{29} + \frac{576059030854044746424320}{190683801}k_{30} - \frac{576059030854044746424320}{190683801}k_{31},$$

$$r_{15} := -\frac{5777383794709213739704218386212223902910709 76}{1455537520992369701623441608028896 93}k_0 + \frac{38994342676877518261046605665243724697989 9136}{1455537520992369701623441608028896 93}k_1$$
$$- \frac{28436863753794178751710742911068035280720 1024}{14555375209923697016234416080288969 3}k_2 + \frac{23371900513179609537931741024881749218319 5904}{14555375209923697016234416080288969 3}k_3$$
$$- \frac{20426661304512879486897058276408292684395 8784}{14555375209923697016234416080288969 3}k_4 + \frac{18098237535036112379745744477041972997606 9632}{14555375209923697016234416080288969 3}k_5$$
$$- \frac{15949248464552147813502440290839840347261 4400}{14555375209923697016234416080288969 3}k_6 + \frac{13913572684817902753531152178475073518055 8336}{14555375209923697016234416080288969 3}k_7$$
$$- \frac{12005129123082760498985501328182048141674 0864}{14555375209923697016234416080288969 3}k_8 + \frac{10240592317521527867693279131367513834816 3072}{14555375209923697016234416080288969 3}k_9$$
$$- \frac{86338518039207496417640219949443743298473 984}{14555375209923697016234416080288969 3}k_{10} + \frac{72043112925773671620161505898310423323652 096}{14555375209923697016234416080288969 3}k_{11}$$
$$- \frac{59828951030446462921950052347269335522385 920}{14555375209923697016234416080288969 3}k_{12} + \frac{50145468776477793634308374381898890313555 968}{14555375209923697016234416080288969 3}k_{13}$$
$$- \frac{43588273326559350446315416451454928314351 616}{14555375209923697016234416080288969 3}k_{14} + \frac{40897809442440987060284075657392619762843 648}{14555375209923697016234416080288969 3}k_{15}$$

$$-\frac{42955861925222786337915243787918244582195200}{145553752099236970162344160802889693}k_{16}+\frac{50780865984879389682162631031032311798038528}{145553752099236970162344160802889693}k_{17}$$

$$-\frac{65521847445400920309068025486770806824632320}{145553752099236970162344160802889693}k_{18}+\frac{88450912673016470567691648142506948199448576}{145553752099236970162344160802889693}k_{19}$$

$$-\frac{120954357456372037081202364100637280542654464}{145553752099236970162344160802889693}k_{20}+\frac{164522623618043879235129804884879526325649408}{145553752099236970162344160802889693}k_{21}$$

$$-\frac{220739419593938362483559138258027962407583744}{145553752099236970162344160802889693}k_{22}+\frac{291270264487621422272510678730059925807693824}{145553752099236970162344160802889693}k_{23}$$

$$-\frac{377850601647604933036669445427582730199105536}{145553752099236970162344160802889693}k_{24}+\frac{482273773863644551181794880666599556095410176}{145553752099236970162344160802889693}k_{25}$$

$$-\frac{606378721801060506038490889931913288422785024}{145553752099236970162344160802889693}k_{26}+\frac{752025380869379181769745921835934191664234496}{145553752099236970162344160802889693}k_{27}$$

$$-\frac{921115675966539272203317511246287289416417280}{145553752099236970162344160802889693}k_{28}+\frac{1117453456888418164994720727898222628872650752}{145553752099236970162344160802889693}k_{29}$$

$$-\frac{1368286209459734662501787199033322451736985600}{145553752099236970162344160802889693}k_{30}+\frac{1869951714602367657515920141303522097465655296}{145553752099236970162344160802889693}k_{31}$$

$$-\frac{3739903429204735315031840282607044194931310592}{145553752099236970162344160802889693}k_{32};$$

$$r_{14} := -\frac{818751761141232181645031751784718093667090841 6}{1892198777290080612110474090437566009}k_0 - \frac{1595330885306224329034562662158941859201572 80}{1892198777290080612110474090437566009}k_1$$

$$+\frac{154394218053931204299148068085218437236530464}{14555375209923697016234416080288969 3}k_2 - \frac{205735030519069475120651298999688868652429251 2}{1892198777290080612110474090437566009}k_3$$

$$+\frac{173240488783546463912099778791680194757446137 6}{1892198777290080612110474090437566009}k_4$$

$$-\frac{140599068080972937049824361091670555485869024 0}{1892198777290080612110474090437566009}k_5$$

$$+\frac{114448865196589456874936379043790772638413683 2}{1892198777290080612110474090437566009}k_6$$

$$-\frac{939844587890153231821926456516923457051665920}{1892198777290080612110474090437566009}k_7$$

$$+\frac{775530716458450573108031044536332857186887680}{1892198777290080612110474090437566009}k_8$$

$$-\frac{641215590606143303960670664662711172566602752}{1892198777290080612110474090437566009}k_9$$

$$+\frac{533598746312304368323837358420614698018730496}{1892198777290080612110474090437566009}k_{10}$$

$$-\frac{349785731147282271348466085925924322345016 32}{14555375209923697016234416080288969 3}k_{11}$$

$$+\frac{410580247182314008263876827317175039683871744}{1892198777290080612110474090437566009}k_{12}$$

$$-\frac{410429903343374999751381395132964252726148096}{1892198777290080612110474090437566009}k_{13}$$

$$+\frac{466571397956138819403384731590732091436726272}{1892198777290080612110474090437566009}k_{14}$$

$$-\frac{594395101150790885129246689338994483319664640}{1892198777290080612110474090437566009}k_{15}$$

$$+\frac{812520028837847841388581651545269952176578560}{1892198777290080612110474090437566009}k_{16}$$

$$-\frac{1142939484077410993309324621624796185836421120}{1892198777290080612110474090437566009}k_{17}$$

$$+\frac{1611122320404854212801419430219237356852150272}{1892198777290080612110474090437566009}k_{18}$$

$$-\frac{2246042283397163284978426120120548188157902848}{1892198777290080612110474090437566009}k_{19}$$

$$+\frac{3080125446429399549359394621911824478685888512}{1892198777290080612110474090437566009}k_{20}$$

$$-\frac{4149118181721195882382697028838574923427086336}{1892198777290080612110474090437566009}k_{21}$$

$$
\begin{aligned}
&+\frac{4224528040138043741855849061334181118056509440}{1455537520992369701623441608028889693}k_{22}\\
&-\frac{7150157392461355980247977668273897149643358208}{18921987772900806121104740904375666009}k_{23}\\
&+\frac{9168214513517564713799636815721714863928508416}{18921987772900806121104740904375666009}k_{24}\\
&-\frac{11592569199135583258269488881518337074348949504}{18921987772900806121104740904375666009}k_{25}\\
&+\frac{14471584786483274457333207749537373737259892736}{18921987772900806121104740904375666009}k_{26}\\
&-\frac{17854724727195931687505586900600532885838233600}{18921987772900806121104740904375666009}k_{27}\\
&+\frac{21793652701843709390979659009425718706904236032}{18921987772900806121104740904375666009}k_{28}\\
&-\frac{26391696071566916452141489425448739574098427904}{18921987772900806121104740904375666009}k_{29}\\
&+\frac{32307970231440982228678836455867430533548998656}{18921987772900806121104740904375666009}k_{30}\\
&-\frac{44140518555118911378175353051670481245245 0140160}{18921987772900806121104740904375666009}k_{31}\\
&+\frac{88281037102378227563507061033409624904900280320}{18921987772900806121104740904375666009}k_{32}.
\end{aligned}
$$

$$
\begin{aligned}
r_{17} := &-\frac{64332988297768504118642912569931385851 0848}{1937556986340307203750267861 0651}k_0 + \frac{14730431296029397027171374560006997 6768512}{1937556986340307203750267861 0651}k_1 \\
&-\frac{1210058973360542406118624016827790540800}{1490428451031005541346359893127}k_2 + \frac{1466508444348165559897402255268056875008}{1937556986340307203750267861 0651}k_3 \\
&-\frac{1379183063693741520231743809249396336 2304}{1937556986340307203750267861 0651}k_4 + \frac{2558390814884655853373829287798228841 2672}{1937556986340307203750267861 0651}k_5 \\
&-\frac{3179324974743198782219756206924513784 6272}{1937556986340307203750267861 0651}k_6 + \frac{3339359015654100987788742150111067268 7104}{1937556986340307203750267861 0651}k_7 \\
&-\frac{32146613512659643567497135901386798530560}{1937556986340307203750267861 0651}k_8 \\
&+\frac{29340222814323358302435599361999339782144}{1937556986340307203750267861 0651}k_9 \\
&-\frac{25727794074444541470428892293017424625664}{1937556986340307203750267861 0651}k_{10} \\
&+\frac{1669909346001985082043437607054018609152}{1490428451031005541346359893127}k_{11} \\
&-\frac{17485837707243561661643418134480882237440}{1937556986340307203750267861 0651}k_{12} \\
&+\frac{13159684454152305001121263037913470861312}{1937556986340307203750267861 0651}k_{13} \\
&-\frac{8780131102450043498347309361207922393088}{1937556986340307203750267861 0651}k_{14} \\
&+\frac{4371095296436930207744006869605743329280}{1937556986340307203750267861 0651}k_{15} \\
&+\frac{568844644476023143190488737005730529 28}{1937556986340307203750267861 0651}k_{16} \\
&-\frac{45005635638585445752450071840751600271 36}{1937556986340307203750267861 0651}k_{17} \\
&+\frac{89611453799683859962186482391606695034 88}{1937556986340307203750267861 0651}k_{18} \\
&-\frac{13443049983427965289224702813189030805504}{1937556986340307203750267861 0651}k_{19} \\
&+\frac{17953498180473373613836035222480375775232}{1937556986340307203750267861 0651}k_{20} \\
&-\frac{22502443030732959711290340551179531976704}{\text{[illegible]}}k_{21}
\end{aligned}
$$

$$
\begin{aligned}
&+\frac{2084817310202454900942859214086690045952}{1490428451031005541346359893127}k_{22}\\
&-\frac{3176959661743064692183014091940334587 0848}{1937556986340307203750267861065 1}k_{23}\\
&+\frac{3652170048921066609954528306837126329 1392}{1937556986340307203750267861065 1}k_{24}\\
&-\frac{4137997088543533927002740748186155116 1344}{1937556986340307203750267861065 1}k_{25}\\
&+\frac{4636788067866044459489253248364853880 4224}{1937556986340307203750267861065 1}k_{26}\\
&-\frac{5151083557995526811655273332640061299 0976}{1937556986340307203750267861065 1}k_{27}\\
&+\frac{5683810697741611163901219912057427538 7392}{1937556986340307203750267861065 1}k_{28}\\
&-\frac{6240823764723557877027930808708252932 5056}{1937556986340307203750267861065 1}k_{29}\\
&+\frac{6846408686177229311916170339825996634 5216}{1937556986340307203750267861065 1}k_{30}\\
&-\frac{8057578529084572181692649402061484038 5536}{1937556986340307203750267861065 1}k_{31}\\
&+\frac{1611515705816914436338529880412296807 71072}{1937556986340307203750267861065 1}k_{32},
\end{aligned}
$$

$$r_{18} := -\frac{720081759589318044553609684653486006474752 0}{2102443085877867346789415656041740 01}k_0 - \frac{6998804769585734064045981619545288517039456}{210244308587786734678941565604174001}k_1$$

$$+\frac{5304400272265360766725319096497657643 19776}{1617263912213744112914935120032 1077}k_2 - \frac{6824456919060819825768983080500742798450336}{210244308587786734678941565604174001}k_3$$

$$+\frac{6745344487584820041562803126194655389563328}{210244308587786734678941565604174001}k_4 - \frac{6631258703671288380471438380101647464928448}{210244308587786734678941565604174001}k_5$$

$$+\frac{6455113775897888780618057957975709314213504}{210244308587786734678941565604174001}k_6$$

$$-\frac{6181235253021515552883083212777760046814720}{210244308587786734678941565604174001}k_7$$

$$+\frac{5758119554980403429819093457588757301356544}{210244308587786734678941565604174001}k_8$$

$$-\frac{5110522418215584299116111920300401752339456}{210244308587786734678941565604174001}k_9$$

$$+\frac{4129576966405346966954698319541915127431680}{210244308587786734678941565604174001}k_{10}$$

$$-\frac{204626082710675433817370075612361538504192}{16172639122137441129149351200321077}k_{11}$$

$$+\frac{4848843006107492838813354871386371462850 56}{210244308587786734678941565604174001}k_{12}$$

$$+\frac{2694985478451938699706852245592271879365632}{210244308587786734678941565604174001}k_{13}$$

$$-\frac{7282123882134533727480605253259587403565056}{210244308587786734678941565604174001}k_{14}$$

$$+\frac{13807628039746943061694296140356476279492608}{210244308587786734678941565604174001}k_{15}$$

$$-\frac{22956294155949443695821205372887579311153152}{210244308587786734678941565604174001}k_{16}$$

$$+\frac{35590428396275037094309238771169584563453952}{210244308587786734678941565604174001}k_{17}$$

$$-\frac{52769844592919353622636487797537224668282880}{210244308587786734678941565604174001}k_{18}$$

$$+\frac{75765966797134442187626316939977444164370432}{210244308587786734678941565604174001}k_{19}$$

$$-\frac{106068830099335348925610826044828633042780160}{210244308587786734678941565604174001}k_{20}$$

$$+\frac{145386924761036033230957881256012322449915904}{210244308587786734678941565604174001}k_{21}$$

$$
\begin{aligned}
&-\frac{1504927932858690469363498577559692636952985 6}{16172639122137441129149351200321077}k_{22}\\
&+\frac{258950098403709600992922925102622957931855872}{2102443085877867346789415656 04174001}k_{23}\\
&-\frac{337618625261985621511998958776201715539509248}{210244308587786734678941565604174001}k_{24}\\
&+\frac{434113669315911983121555959326471394635022336}{210244308587786734678941565604174001}k_{25}\\
&-\frac{551043106399733822502402833747296646861946880}{210244308587786734678941565604174001}k_{26}\\
&+\frac{691107750712581020350483488820461979164999680}{210244308587786734678941565604174001}k_{27}\\
&-\frac{857194293183352945392011830903591609091227648}{210244308587786734678941565604174001}k_{28}\\
&+\frac{1055076115669848335079415672711933971734724608}{210244308587786734678941565604174001}k_{29}\\
&-\frac{1316548498187790654058570513970701799812759552}{210244308587786734678941565604174001}k_{30}\\
&+\frac{1839493263223675292016880196488237455968829440}{210244308587786734678941565604174001}k_{31}\\
&-\frac{3678986526447350584033760392976474911937658880}{210244308587786734678941565604174001}k_{32},
\end{aligned}
$$

$$
\begin{aligned}
r_{19} := & -\frac{846331556576961593459224129208633418441519104}{1892198777290080612110474090437566009}k_0 + \frac{376685845259126425679159929193982114613026400}{1892198777290080612110474090437566009}k_1 \\
& -\frac{159493323008700525201945318772721189753787 20}{145553752099236970162344160802889693}k_2 + \frac{157457424374236669142626570672887235024901536}{1892198777290080612110474090437566009}k_3 \\
& -\frac{142757330201442573398181736488684165478953920}{1892198777290080612110474090437566009}k_4 + \frac{134052406517472853730290982190011910207067840}{1892198777290080612110474090437566009}k_5 \\
& -\frac{124259776276963153692996588503959035499995776}{1892198777290080612110474090437566009}k_6 \\
& +\frac{112921269720863783677292219739652208623232512}{1892198777290080612110474090437566009}k_7 \\
& -\frac{100743677662559060822473284906275626382305280}{1892198777290080612110474090437566009}k_8 \\
& +\frac{88244241456936317146220738290478244155806720}{1892198777290080612110474090437566009}k_9 \\
& -\frac{75651309378062869097629433955447348840395264}{1892198777290080612110474090437566009}k_{10} \\
& +\frac{4849031507067062092856558657652005348873728}{145553752099236970162344160802889693}k_{11} \\
& -\frac{50417462117028014651397187179764440190501888}{1892198777290080612110474090437566009}k_{12} \\
& +\frac{37790993654245889191778157000988234520083456}{1892198777290080612110474090437566009}k_{13} \\
& -\frac{25154311114530973778498491223713198830811136}{1892198777290080612110474090437566009}k_{14} \\
& +\frac{12501854301022655068735728536552256371814400}{1892198777290080612110474090437566009}k_{15} \\
& +\frac{174587763380230960804941100576736416768000}{1892198777290080612110474090437566009}k_{16} \\
& -\frac{12886880971902362237482557821386000286842880}{1892198777290080612110474090437566009}k_{17} \\
& +\frac{25651762948709973544400826181308544276037632}{1892198777290080612110474090437566009}k_{18} \\
& -\frac{38492215473544975648972869943687091588956160}{1892198777290080612110474090437566009}k_{19} \\
& +\frac{51438870966597824196638913979421144158044160}{1892198777290080612110474090437566009}k_{20} \\
& -\frac{64531272182658442150283331484088449896808448}{1892198777290080612110474090437566009}k_{21}
\end{aligned}
$$

$$+\frac{5986060837788585057029513883773431389356032}{145553752099236970162344160802889693}k_{22}$$
$$-\frac{9136108980523126207082724786326428210744524 8}{18921987772900806121104740904375660 09}k_{23}$$
$$+\frac{10522816899184436284113640744276695736070963 2}{18921987772900806121104740904375660 09}k_{24}$$
$$-\frac{11950007923428538318358540556208311815634944 0}{18921987772900806121104740904375660 09}k_{25}$$
$$+\frac{13426596929423222106751000013892512639027773 44}{18921987772900806121104740904375660 09}k_{26}$$
$$-\frac{14962218043691792337205674759592194121190604 8}{18921987772900806121104740904375660 09}k_{27}$$
$$+\frac{16567943893468598028973378386096979110264832 0}{18921987772900806121104740904375660 09}k_{28}$$
$$-\frac{18265919733222329945868738954214409561282969 6}{18921987772900806121104740904375660 09}k_{29}$$
$$+\frac{20148395552929914313019413405557130936188928 0}{18921987772900806121104740904375660 09}k_{30}$$
$$-\frac{23913347192345083047320762308242573686000844 8}{18921987772900806121104740904375660 09}k_{31}$$
$$+\frac{47826694384690166094641524616485147372001689 6}{18921987772900806121104740904375660 09}k_{32};$$

$$
\begin{aligned}
r_{20} :=\; & \frac{1078859109831985493644073609782886992384}{1892198777290080612110474090437566009}k_0 + \frac{4499676737645056443877707160369439025260}{1892198777290080612110474090437566009}k_1 \\
& - \frac{40663687438224428879871782230115978832 4}{145553752099236970162344160802889693}k_2 + \frac{5081939553175629806444613010575260341900}{1892198777290080612110474090437566009}k_3 \\
& - \frac{4709782744231551655955564510815646508184}{1892198777290080612110474090437566009}k_4 + \frac{4317649540934262446189484869616220726928}{1892198777290080612110474090437566009}k_5 \\
& - \frac{3924794600960911189951601678871639244640}{1892198777290080612110474090437566009}k_6 + \frac{3532247300518338232573402434653887273472}{1892198777290080612110474090437566009}k_7 \\
& - \frac{3139765231809650633289399363326082586624}{1892198777290080612110474090437566009}k_8 + \frac{2747293623660562738263020611062084637696}{1892198777290080612110474090437566009}k_9 \\
& - \frac{2354822949354024558118710138817669047808}{1892198777290080612110474090437566009}k_{10} + \frac{150950166734930474975341994482256099840}{145553752099236970162344160802889693}k_{11} \\
& - \frac{1569881213199230909496003432450828372992}{1892198777290080612110474090437566009}k_{12} + \frac{1177409985069016148647654302934124985344}{1892198777290080612110474090437566009}k_{13} \\
& - \frac{784938328109830133555525743770575190016}{1892198777290080612110474090437566009}k_{14} \\
& + \frac{392466008874649421729581783440263753728}{1892198777290080612110474090437566009}k_{15} \\
& + \frac{73173831237067801265026526839193 60}{1892198777290080612110474090437566009}k_{16} \\
& - \frac{392482148874754034898390804238201520128}{1892198777290080612110474090437566009}k_{17} \\
& + \frac{784959188374318952231532230179666395136}{1892198777290080612110474090437566009}k_{18} \\
& - \frac{1177439400853839509168059996346859618304}{1892198777290080612110474090437566009}k_{19} \\
& + \frac{1569924072548545823008794206264513134592}{1892198777290080612110474090437566009}k_{20} \\
& - \frac{1962414863846032624142286378132182958080}{1892198777290080612110474090437566009}k_{21} \\
& + \frac{181147219062231418506894629602025209856}{145553752099236970162344160802889693}k_{22}
\end{aligned}
$$

$$-\frac{2747423529387971157406848123153438048256}{1892198777290080612110474090437566009}k_{23}$$
$$+\frac{3139946846873401163761226327938994274304}{1892198777290080612110474090437566009}k_{24}$$
$$-\frac{3532487158686654513314341962121808084992}{1892198777290080612110474090437566009}k_{25}$$
$$+\frac{3925048200848256111218475758859131617280}{1892198777290080612110474090437566009}k_{26}$$
$$-\frac{4317634001914486088109437105736875737088}{1892198777290080612110474090437566009}k_{27}$$
$$+\frac{4710249175513135025590092699198268768256}{1892198777290080612110474090437566009}k_{28}$$
$$-\frac{5102902948899784086866175932129767358464}{1892198777290080612110474090437566009}k_{29}$$
$$+\frac{5495633921862433395733114444001477066752}{1892198777290080612110474090437566009}k_{30}$$
$$-\frac{6281095867787732013466991467744896483328}{1892198777290080612110474090437566009}k_{31}$$
$$+\frac{12562191735575464026933982935489792966656}{1892198777290080612110474090437566009}k_{32},$$

$$
\begin{aligned}
r_{21} :=& -\frac{7750445741315666895918359690548182249472}{1892198777290080612110474090437566009}k_0 + \frac{24220143775312900300632315978844443416864}{6307329257633602040368246968125220003}k_1 \\
& -\frac{1738882140813988851274850817930201112 00}{4851791736641232338744805360096323 1}k_2 + \frac{6297237498249787169137694842496337310880}{1892198777290080612110474090437566009}k_3 \\
& -\frac{2152901712357599797067525114372427072 64}{7008143619592891155964718853472466 7}k_4 + \frac{2192770263673625390976976720400852998 4}{7786826243992101284405243170524963}k_5 \\
& -\frac{4844028855842384549488996906287563499136}{1892198777290080612110474090437566009}k_6 + \frac{1453208656801679115994283739232375591424}{6307329257633602040368246968125220003}k_7 \\
& -\frac{1291741028294309961352031286542241658880}{6307329257633602040368246968125220003}k_8 + \frac{3390820199402965390086824038297507251200}{1892198777290080612110474090437566009}k_9 \\
& -\frac{968805771340934266889385663715470591488}{6307329257633602040368246968125220003}k_{10} + \frac{621029340720122256796636994792654443 52}{4851791736641232338744805360096323 1}k_{11} \\
& -\frac{1937611543862008717801004653814447653888}{1892198777290080612110474090437566009}k_{12} + \frac{1614676288159559122652403245306683504 64}{2102443085877867346789415656041740 01}k_{13} \\
& -\frac{1076450861668247392907131055890829373 44}{2102443085877867346789415656041740 01}k_{14} \\
& +\frac{4844028927141762518987530512189693747 20}{1892198777290080612110474090437566009}k_{15} \\
& -\frac{385013904338540722886103440588 8}{6307329257633602040368246968125220003}k_{16} \\
& -\frac{1614676223837973837669629575623463731 20}{6307329257633602040368246968125220003}k_{17} \\
& +\frac{9688057421784554347652554791590397870 08}{1892198777290080612110474090437566009}k_{18} \\
& -\frac{4844028706042013117918804419305985802 24}{6307329257633602040368246968125220003}k_{19} \\
& +\frac{6458704912241822468781445759750016860 16}{6307329257633602040368246968125220003}k_{20} \\
& -\frac{2422014324531738330746397181737959686144}{1892198777290080612110474090437566009}k_{21} \\
& +\frac{2484117230868160602924272305417800908 8}{161726391221374411291493512003210 77}k_{22}
\end{aligned}
$$

$$
\begin{aligned}
&-\frac{37675777498698233289145736680714692 1984}{2102443085877867346789415656041 74001}k_{23}\\
&+\frac{387522276179113522842425379640711105 7408}{1892198777290080612110474090437566009}k_{24}\\
&-\frac{1453208504019684655763100085135207825408}{630732925763360204036824696812522003}k_{25}\\
&+\frac{1614676071637458574404128160319563366400}{630732925763360204036824696812522003}k_{26}\\
&-\frac{5328430857078332966992982826454813769728}{1892198777290080612110474090437566009}k_{27}\\
&+\frac{1937611140095688696613755582220613976064}{630732925763360204036824696812522003}k_{28}\\
&-\frac{2099078622666285291392526383705156485120}{630732925763360204036824696812522003}k_{29}\\
&+\frac{6781638084716758981489948524062295719936}{1892198777290080612110474090437566009}k_{30}\\
&-\frac{2870534265982431554116550840724425277 44}{700814361959289115596471885347246 67}k_{31}\\
&+\frac{5741068531964863108233101681448850554 88}{700814361959289115596471885347246 67}k_{32},
\end{aligned}
$$

$$
\begin{aligned}
r_{22} := & -\frac{7330595268122946758044910723058938685685760}{145553752099236970162344160802889693}k_0 + \frac{9006593137467592173065483152486192627712000}{145553752099236970162344160802889693}k_1 \\
& - \frac{1037914273806997571223135875943027397269913 6}{145553752099236970162344160802889693}k_2 + \frac{11447140100956022325813815640931025173479424}{145553752099236970162344160802889693}k_3 \\
& - \frac{12210136358542666643094706709896919005265920}{145553752099236970162344160802889693}k_4 + \frac{12667964437324526499003420546606850309357568}{145553752099236970162344160802889693}k_5 \\
& - \frac{12820567619976158014667894839843376314449920}{145553752099236970162344160802889693}k_6 + \frac{12667928623549064239962099890421036386091008}{145553752099236970162344160802889693}k_7 \\
& - \frac{12210042926973817754320116320247206096404480}{145553752099236970162344160802889693}k_8 + \frac{11446909727181089695980728439132181434990592}{145553752099236970162344160802889693}k_9 \\
& - \frac{10378529252103120978703064202643082366156800}{145553752099236970162344160802889693}k_{10} + \frac{9004902143711410259623970831102950246449152}{145553752099236970162344160802889693}k_{11} \\
& - \frac{7326029453415436983893725519258810924924928}{145553752099236970162344160802889693}k_{12} + \frac{5341912832285938079558450269256713313976320}{145553752099236970162344160802889693}k_{13} \\
& - \frac{3052554837953837399143495350450541960364032}{145553752099236970162344160802889693}k_{14} \\
& + \frac{457959373624724012185649114651507210321920}{145553752099236970162344160802889693}k_{15} \\
& + \frac{2441867700628816200519995307346254708604928}{145553752099236970162344160802889693}k_{16} \\
& - \frac{5646917742386191009346909071518117451005952}{145553752099236970162344160802889693}k_{17} \\
& + \frac{9157178250108114291442578304123637510176768}{145553752099236970162344160802889693}k_{18} \\
& - \frac{12972631511676105821867427090785057634254848}{145553752099236970162344160802889693}k_{19} \\
& + \frac{17093252977057450133634166920944605064069120}{145553752099236970162344160802889693}k_{20} \\
& - \frac{21519009378845941910931166701457110097461248}{145553752099236970162344160802889693}k_{21} \\
& + \frac{26249856640493667160119769663408643413901312}{145553752099236970162344160802889693}k_{22}
\end{aligned}
$$

$$-\frac{31285737610373723282181200189371267095199744}{1455537520992369701623441608028896 93}k_{23}$$

$$+\frac{36626579656463088928460136994542787282075648}{145553752099236970162344160802889693}k_{24}$$

$$-\frac{42272292191224065403399105528669284346101760}{145553752099236970162344160802889693}k_{25}$$

$$+\frac{48222764056882861161254508819533198092402688}{145553752099236970162344160802889693}k_{26}$$

$$-\frac{54477858295986760799528689159929498165051392}{145553752099236970162344160802889693}k_{27}$$

$$+\frac{61037386351725201202215867372678213886017536}{145553752099236970162344160802889693}k_{28}$$

$$-\frac{67900971110477056137304485103419403899240448}{145553752099236970162344160802889693}k_{29}$$

$$+\frac{75372669275255740137195981870145542496976896}{145553752099236970162344160802889693}k_{30}$$

$$-\frac{90316065604813108136978975403597819692449792}{145553752099236970162344160802889693}k_{31}$$

$$+\frac{180632131209626216273957950807195639384899584}{145553752099236970162344160802889693}k_{32};$$

$$
\begin{aligned}
r_{23} := {} & \frac{490499379497699109982390499235594189716684 8}{1892198777290080612110474090437566009}k_0 - \frac{8211303562962320504434873921405623355703 68}{1892198777290080612110474090437566009}k_1 \\
& - \frac{827985679637822784681700055667561357280 0}{145553752099236970162344160802889693}k_2 + \frac{269327618836431288290571063581891599820224}{1892198777290080612110474090437566009}k_3 \\
& - \frac{271729813216736475220990849337008590235264}{1892198777290080612110474090437566009}k_4 + \frac{249367267213180302611116340318864818363520}{1892198777290080612110474090437566009}k_5 \\
& - \frac{225649489483342663184680707166035600733952}{1892198777290080612110474090437566009}k_6 + \frac{202714277096662969776878653545620790993920}{1892198777290080612110474090437566009}k_7 \\
& - \frac{180119193510954322575954518104304140251136}{1892198777290080612110474090437566009}k_8 + \frac{157598066119643530302179040552462253287424}{1892198777290080612110474090437566009}k_9 \\
& - \frac{135083529689461825490979043654947111156736}{1892198777290080612110474090437566009}k_{10} + \frac{8659093361664868199144013300859307058176}{145553752099236970162344160802889693}k_{11} \\
& - \frac{90051646126494768276572685726275511744512}{1892198777290080612110474090437566009}k_{12} + \frac{67533092422513980993802478813703822522368}{1892198777290080612110474090437566009}k_{13} \\
& - \frac{45011427064271071866336070367548073046016}{1892198777290080612110474090437566009}k_{14} \\
& + \frac{22484954875052346128059614752663873609728}{1892198777290080612110474090437566009}k_{15} \\
& + \frac{48828618001527661737477211041575813120}{1892198777290080612110474090437566009}k_{16} \\
& - \frac{22593544739321788598541316502269149773824}{1892198777290080612110474090437566009}k_{17} \\
& + \frac{45154305249628187288757857213057840840704}{1892198777290080612110474090437566009}k_{18} \\
& - \frac{67738135284274891919957102500686651719680}{1892198777290080612110474090437566009}k_{19}
\end{aligned}
$$

$$+\frac{9035440954860440500928572446965233929420 8}{1892198777290080612110474090437566009}k_{20}-\frac{1130152477164545512017366207037814063759 36}{1892198777290080612110474090437566009}k_{21}$$

$$+\frac{104412162050043487546193353383921078763 52}{145553752099236970162344160802889693}k_{22}-\frac{158534462155308165933053765885443805544448}{1892198777290080612110474090437566009}k_{23}$$

$$+\frac{181432806517898598279257504999621181571072}{1892198777290080612110474090437566009}k_{24}-\frac{204455623466237187949433368235556009410560}{1892198777290080612110474090437566009}k_{25}$$

$$+\frac{227630597162990649137095970448532840644608}{1892198777290080612110474090437566009}k_{26}-\frac{250987774353157354529477143263658618388480}{1892198777290080612110474090437566009}k_{27}$$

$$+\frac{274561926946398993801243151845685068824576}{1892198777290080612110474090437566009}k_{28}-\frac{298422598762038946301726526524115644317696}{1892198777290080612110474090437566009}k_{29}$$

$$+\frac{322856309022475525259644633395354469924864}{1892198777290080612110474090437566009}k_{30}-\frac{371723729543348683175480847137832121139200}{1892198777290080612110474090437566009}k_{31}$$

$$+\frac{743447459086697366350961694275664242278400}{1892198777290080612110474090437566009}k_{32},$$

$$
\begin{aligned}
r_{24} := {} & \frac{145773107328387787239495588762715617161641984}{630732925763360204036824696812522003}k_0 - \frac{134765685349588570290149688612291480739397632}{630732925763360204036824696812522003}k_1 \\
& + \frac{9615713480120246784518403483387207759478784}{48517917366412323387448053600963231}k_2 - \frac{116408011101702803812430720217175956637859840}{630732925763360204036824696812522003}k_3 \\
& + \frac{108747836915362164721541759501412687669788672}{630732925763360204036824696812522003}k_4 - \frac{101730916482220209658193550268140666410008576}{630732925763360204036824696812522003}k_5 \\
& + \frac{95039435947755773606098145007971966215192576}{630732925763360204036824696812522003}k_6 - \frac{88346668352253490169654967760481370656997376}{630732925763360204036824696812522003}k_7 \\
& + \frac{81323025266904993979063586830755995784839168}{630732925763360204036824696812522003}k_8 - \frac{73638123063840656748351293497328834686484480}{630732925763360204036824696812522003}k_9 \\
& + \frac{64961419707995851679177999477506380815138816}{630732925763360204036824696812522003}k_{10} - \frac{4227876781491018815427531934689850116800512}{48517917366412323387448053600963231}k_{11} \\
& + \frac{43310633939484060954607575895935643346796544}{630732925763360204036824696812522003}k_{12} - \frac{29675851548640485090500617540300569985417216}{630732925763360204036824696812522003}k_{13} \\
& + \frac{13728000703916946284035196433024650450567168}{630732925763360204036824696812522003}k_{14} \\
& + \frac{4862634610354909208900888129411935899222016}{630732925763360204036824696812522003}k_{15} \\
& - \frac{26425284310288950835265579508058705799151616}{630732925763360204036824696812522003}k_{16} \\
& + \frac{51288487200928294427805450637245641295134720}{630732925763360204036824696812522003}k_{17} \\
& - \frac{79779823591639442260286655377655384600215552}{630732925763360204036824696812522003}k_{18} \\
& + \frac{112225579759759682567972165651870111449481216}{630732925763360204036824696812522003}k_{19} \\
& - \frac{148950344015546216832872275886834932520058880}{630732925763360204036824696812522003}k_{20}
\end{aligned}
$$

$$
\begin{aligned}
&+\frac{190276540325411647454137656363868830540759040}{630732925763360204036824696812522003}k_{21}\\
&-\frac{18194146879850300162757304070941297647026176}{48517917366412323387448053600963231}k_{22}\\
&+\frac{288008947013086673598382477846067281621680128}{630732925763360204036824696812522003}k_{23}\\
&-\frac{345044307361772458647558631531936780201754624}{630732925763360204036824696812522003}k_{24}\\
&+\frac{407938190019062363129678334131770787664232448}{630732925763360204036824696812522003}k_{25}\\
&-\frac{476993691978249439758555477146373498932297728}{630732925763360204036824696812522003}k_{26}\\
&+\frac{552507512160275363549537826612488007195295744}{630732925763360204036824696812522003}k_{27}\\
&-\frac{634763553341379054121040897638735202174894080}{630732925763360204036824696812522003}k_{28}\\
&+\frac{724329621074038350709895720442356863314427904}{630732925763360204036824696812522003}k_{29}\\
&-\frac{828515741909808859333454046800727456773832704}{630732925763360204036824696812522003}k_{30}\\
&+\frac{1036887983581349876580570699517468643692642304}{630732925763360204036824696812522003}k_{31}\\
&-\frac{2073775967162699753161141399034937287385284608}{630732925763360204036824696812522003}k_{32},
\end{aligned}
$$

$$
\begin{aligned}
r_{25} :=\ & \frac{30369535433400425692686935693374976}{523140386311882945012572322487577}k_0 - \frac{30548727162750800714204873557216492}{523140386311882945012572322487577}k_1 \\
& + \frac{23411562189455259251213381431397 48}{40241568177837149616351717114429}k_2 - \frac{30024872588068556611637930492565772}{523140386311882945012572322487577}k_3 \\
& + \frac{29232599827905537079386264036365336}{523140386311882945012572322487577}k_4 - \frac{27842804442099501942911262650939152}{523140386311882945012572322487577}k_5 \\
& + \frac{25459204842448720792168607543575520}{523140386311882945012572322487577}k_6 - \frac{21407452619120992442617832457534592}{523140386311882945012572322487577}k_7 \\
& + \frac{14567206920061526753831976512856320}{523140386311882945012572322487577}k_8 - \frac{30993383594072164046036526456 26368}{523140386311882945012572322487577}k_9 \\
& - \frac{15984819201215476418236924769695744}{523140386311882945012572322487577}k_{10} + \frac{3653167927184696226890820477117440}{40241568177837149616351717114429}k_{11} \\
& - \frac{99060197958369474007053504906299392}{523140386311882945012572322487577}k_{12} + \frac{182687880879359203066337060250460160}{523140386311882945012572322487577}k_{13} \\
& - \frac{316952615901078579346220447703138304}{523140386311882945012572322487577}k_{14} \\
& + \frac{530197867579982217365618357622456320}{523140386311882945012572322487577}k_{15} \\
& - \frac{864958160190631396682799005983473664}{523140386311882945012572322487577}k_{16} \\
& + \frac{1383923756297083969736486490216366080}{523140386311882945012572322487577}k_{17} \\
& - \frac{2177695291120799781815050956642254848}{523140386311882945012572322487577}k_{18} \\
& + \frac{3374461682473263310402763715600416768}{523140386311882945012572322487577}k_{19} \\
& - \frac{5151536628577199013849272625507991552}{523140386311882945012572322487577}k_{20}
\end{aligned}
$$

$$+\frac{7748434988578592750116470679442849792}{523140386311882945012572322487577}k_{21}$$
$$-\frac{883146802636934355272972131593551872}{40241568177837149616351717114429}k_{22}$$
$$+\frac{16755082077855360812570820857829883904}{523140386311882945012572322487577}k_{23}$$
$$-\frac{24080278412759568541138590877071179776}{523140386311882945012572322487577}k_{24}$$
$$+\frac{34077701242270576134585448479248777216}{523140386311882945012572322487577}k_{25}$$
$$-\frac{47480839583288513519265530714712506368}{523140386311882945012572322487577}k_{26}$$
$$+\frac{65136275473935535166839023743392514048}{523140386311882945012572322487577}k_{27}$$
$$-\frac{88116776994777844639578211944379580416}{523140386311882945012572322487577}k_{28}$$
$$+\frac{118567880306947849062027664075855265792}{523140386311882945012572322487577}k_{29}$$
$$-\frac{163960187201773243383897644068308189184}{523140386311882945012572322487577}k_{30}$$
$$+\frac{254744800991424032027637604053214035968}{523140386311882945012572322487577}k_{31}$$
$$-\frac{509489601982848064055275208106428071936}{523140386311882945012572322487577}k_{32}$$

$$
\begin{aligned}
r_{26} :=& -\frac{532576647580652673203087099587540161536}{630732925763360204036824696812522003}k_0 - \frac{178926184230413258847165142666539260008}{778682624399210128440524317052496\,3}k_1 \\
&+\frac{445058010135100812633179906558722216\,88}{161726391221374411291493512003210\,77}k_2 - \frac{167265897884983332223728083614511266\,8200}{630732925763360204036824696812522003}k_3 \\
&+\frac{516858628044770681060100011005773225232}{210244308587786734678941565604174001}k_4 - \frac{473827854412874945611124169162681013808}{210244308587786734678941565604174001}k_5 \\
&+\frac{129213801158971811752995674310317610659\,2}{630732925763360204036824696812522003}k_6 - \frac{387633468742758261653150900325437399296}{210244308587786734678941565604174001}k_7 \\
&+\frac{114853965527070992712320085170737263616}{70081436195928911559647188534724667}k_8 - \frac{904474643717864850160291642596684798464}{630732925763360204036824696812522003}k_9 \\
&+\frac{861404361513575297541161732594867284\,48}{70081436195928911559647188534724667}k_{10} - \frac{165654658791219482410303720654127372\,80}{161726391221374411291493512003210\,77}k_{11} \\
&+\frac{516842352657515900808283672661766988288}{630732925763360204036824696812522003}k_{12} - \frac{129210479985959956144078257625580354048}{210244308587786734678941565604174001}k_{13} \\
&+\frac{861401255921257360753811445675741153\,28}{210244308587786734678941565604174001}k_{14} \\
&-\frac{129209081193113319790421238556112648192}{630732925763360204036824696812522003}k_{15} \\
&-\frac{855922860290834750886914685644800}{210244308587786734678941565604174001}k_{16} \\
&+\frac{478573129402549897911232642597730713\,6}{233604787319763038532157295115748\,89}k_{17} \\
&-\frac{258427696931956483440618642038062481408}{630732925763360204036824696812522003}k_{18} \\
&+\frac{430713069303944933995596672230040535\,04}{70081436195928911559647188534724667}k_{19} \\
&-\frac{172285797542331673982701672178729353216}{210244308587786734678941565604174001}k_{20}
\end{aligned}
$$

$$
\begin{aligned}
&+\frac{646075170355437997575776455063803658240}{630732925763360204036824696812522003}k_{21}\\
&-\frac{198793800780079948390524365137030676 48}{161726391221374411291493512003210 77}k_{22}\\
&+\frac{301506743245555160679099524221938171904}{210244308587786734678941565604174001}k_{23}\\
&-\frac{1033749421566836849380460077081460473856}{630732925763360204036824696812522003}k_{24}\\
&+\frac{387661525703268081381862327465684238336}{210244308587786734678941565604174001}k_{25}\\
&-\frac{143580778607988633398265251857243832320}{70081436195928911559647188534724667}k_{26}\\
&+\frac{1421478126450221242462251569114767163392}{630732925763360204036824696812522003}k_{27}\\
&-\frac{57434798267534328613775401875081789440}{23360478731976303853215729511574889}k_{28}\\
&+\frac{560004842049606198832984011105131036672}{210244308587786734678941565604174001}k_{29}\\
&-\frac{1809316591378582508060632018355065192448}{630732925763360204036824696812522003}k_{30}\\
&+\frac{689306907279370110394663996144803119104}{210244308587786734678941565604174001}k_{31}\\
&-\frac{1378613814558740220789327992289606238208}{210244308587786734678941565604174001}k_{32},
\end{aligned}
$$

$$r_{27} := -\frac{24406880405606297118438480262408669696 0}{1892198777290080612110474090437566009}k_0 + \frac{2455219929734201447926191708845455690 64}{1892198777290080612110474090437566009}k_1$$
$$-\frac{1881800278086134037208028150823666162 4}{145553752099236970162344160802889693}k_2 + \frac{2413861552176701643799340964480717692 72}{1892198777290080612110474090437566009}k_3$$
$$-\frac{2351090915845917965195369254016698644 64}{1892198777290080612110474090437566009}k_4 + \frac{2241070607416089311648325539964816101 60}{1892198777290080612110474090437566009}k_5$$
$$-\frac{2052602774317986263428479670114743457 60}{1892198777290080612110474090437566009}k_6 + \frac{1732681066193649706869116867963059980 80}{1892198777290080612110474090437566009}k_7$$
$$-\frac{1193432281312004147424332695070405724 16}{1892198777290080612110474090437566009}k_8 + \frac{2909757541557974574955026251808751616 0}{1892198777290080612110474090437566009}k_9$$
$$+\frac{1207801240520064028991346633155869760 00}{1892198777290080612110474090437566009}k_{10} - \frac{2828034639818589991777034609433594649 6}{145553752099236970162344160802889693}k_{11}$$
$$+\frac{7706381502376525198512510856246173025 28}{1892198777290080612110474090437566009}k_{12} - \frac{1422168184083610334083018337916074206 720}{1892198777290080612110474090437566009}k_{13}$$
$$+\frac{2464519536666052652128858971369315666 944}{1892198777290080612110474090437566009}k_{14} - \frac{4113260258774340391564396738165787318 272}{1892198777290080612110474090437566009}k_{15}$$
$$+\frac{6689185183531792382450120946849544318 976}{1892198777290080612110474090437566009}k_{16} - \frac{1066030848386648766888065747776218859 1104}{1892198777290080612110474090437566009}k_{17}$$
$$+\frac{1669465587426702304477074103857056340 3776}{1892198777290080612110474090437566009}k_{18} - \frac{2572317799166923802035329488988227121 9712}{1892198777290080612110474090437566009}k_{19}$$
$$+\frac{3901007743552802242651904914470960732 5696}{1892198777290080612110474090437566009}k_{20} - \frac{5822571270774631671040273856830049891 1232}{1892198777290080612110474090437566009}k_{21}$$
$$+\frac{6578154782971990678785047592218031030 272}{145553752099236970162344160802889693}k_{22} - \frac{1235628993403557934624469131459032179 01568}{1892198777290080612110474090437566009}k_{23}$$
$$+\frac{1756313668798175031092425529965414480 60928}{1892198777290080612110474090437566009}k_{24} - \frac{2455944746350339198188079301883341323 63264}{1892198777290080612110474090437566009}k_{25}$$
$$+\frac{3379141953183566984624086788166802777 04704}{1892198777290080612110474090437566009}k_{26} - \frac{4576031059629507832222622377493197304 62720}{1892198777290080612110474090437566009}k_{27}$$

$$+\frac{610774992243607696902489655398674015977472}{1892198777290080612110474090437566009}k_{28}-\frac{809657425509910485111333028589586184929280}{1892198777290080612110474090437566009}k_{29}$$

$$+\frac{1099960952747505022377408312863614120755200}{1892198777290080612110474090437566009}k_{30}-\frac{1680568007222694096909558881411669992407040}{1892198777290080612110474090437566009}k_{31}$$

$$+\frac{3361136014445388193819117762823339984814080}{1892198777290080612110474090437566009}k_{32}.$$

Proof. It follows from (6)-(11) that

$$a_1+2a_2+3a_3+4a_4+6a_6+12a_{12}=24b_1, \tag{17}$$

$$a_1+a_2+a_3+a_4+a_6+a_{12}=32, \tag{18}$$

$$-\frac{a_1}{6}-\frac{a_2}{3}-\frac{a_3}{6}-\frac{2a_4}{3}-\frac{a_6}{3}-\frac{2a_{12}}{3}=-b_1-b_5.$$

Now we will use p-k parametrization of Alaca, Alaca and Williams, see [15] :

$$p(q):=\frac{\varphi^2(q)-\varphi^2(q^3)}{2\varphi^2(q^3)},\ k(q):=\frac{\varphi^3(q^3)}{\varphi(q)}, \tag{19}$$

where the theta function $\varphi(q)$ is defined by

$$\varphi(q)=\sum_{-\infty}^{\infty}q^{n^2}.$$

Setting x = p in (12), and multiplying both sides by k^{16} we obtain

$$\begin{aligned}&\frac{k^{16}}{2^{b_1+b_5}}p^{b_1}(1-p)^{b_2}(1+p)^{b_3}(1+2p)^{b_4}(2+p)^{b_5}\\&=\big(k_0+k_1p+k_2p^2+k_3p^3+k_4p^4+k_5p^5+k_6p^6+k_7p^7+k_8p^8+k_9p^9+k_{10}p^{10}\\&\quad+k_{11}p^{11}+k_{12}p^{12}+k_{13}p^{13}+k_{14}p^{14}+k_{15}p^{15}+k_{16}p^{16}+k_{17}p^{17}+k_{18}p^{18}+k_{19}p^{19}\\&\quad+k_{20}p^{20}+k_{21}p^{21}+k_{22}p^{22}+k_{23}p^{23}+k_{24}p^{24}+k_{25}p^{25}+k_{26}p^{26}+k_{27}p^{27}+k_{28}p^{28}\\&\quad+k_{29}p^{29}+k_{30}p^{30}+k_{31}p^{31}+k_{32}p^{32}\big)k^{16}.\end{aligned}$$

Alaca, Alaca and Williams [16] have established the following representations in terms of p and k:

$$\eta(q)=2^{-1/6}p^{1/24}(1-p)^{1/2}(1+p)^{1/6}(1+2p)^{1/8}(2+p)^{1/8}k^{1/2}, \tag{20}$$

$$\eta(q^2)=2^{-1/3}p^{1/12}(1-p)^{1/4}(1+p)^{1/12}(1+2p)^{1/4}(2+p)^{1/4}k^{1/2}, \tag{21}$$

$$\eta(q^3)=2^{-1/6}p^{1/8}(1-p)^{1/6}(1+p)^{1/2}(1+2p)^{1/24}(2+p)^{1/24}k^{1/2}, \tag{22}$$

$$\eta(q^4)=2^{-2/3}p^{1/6}(1-p)^{1/8}(1+p)^{1/24}(1+2p)^{1/8}(2+p)^{1/2}k^{1/2}, \tag{23}$$

$$\eta(q^6)=2^{-1/3}p^{1/4}(1-p)^{1/12}(1+p)^{1/4}(1+2p)^{1/12}(2+p)^{1/12}k^{1/2}, \tag{24}$$

$$\eta(q^{12})=2^{-2/3}p^{1/2}(1-p)^{1/24}(1+p)^{1/8}(1+2p)^{1/24}(2+p)^{1/6}k^{1/2}, \tag{25}$$

$$\begin{aligned}E_6(q)&:=1-504\sum_{n=1}^{\infty}\sigma_5(n)q^n\\&=\big(1-246p-5532p^2-38614p^3-135369p^4-276084p^5-348024p^6\\&\quad-276084p^7-135369p^8-38614p^9-5532p^{10}-246p^{11}+p^{12}\big)k^6,\end{aligned}$$

$$\begin{aligned}E_4(q)&:=1+240\sum_{n=1}^{\infty}\sigma_3(n)q^n\\&=\big(1+124p+964p^2+2788p^3+3910p^4+2788p^5+964p^6+124p^7+p^8\big)k^4.\end{aligned}$$

Therefore, since

$$E_{16}(q)=\frac{2000}{3617}E_6^2(q)E_4(q)+\frac{1617}{3617}E_4^4(q)$$

we immediately obtain:

$$
\begin{aligned}
E_{16}(q)=\Big(&p^{32}+\frac{66032}{3617}p^{31}+\frac{134229104}{3617}p^{30}+\frac{31279681360}{3617}p^{29}+\frac{1532766571768}{3617}p^{28}\\
&+\frac{33608470215088}{3617}p^{27}+\frac{431880430212432}{3617}p^{26}+\frac{3705935707430928}{3617}p^{25}\\
&+\frac{22884510224189340}{3617}p^{24}+\frac{106630640734839600}{3617}p^{23}+\frac{387088285203107184}{3617}p^{22}\\
&+\frac{1119620321590028304}{3617}p^{21}+\frac{2622064678835063496}{3617}p^{20}+\frac{5029685780950817520}{3617}p^{19}\\
&+\frac{7966990925710242000}{3617}p^{18}+\frac{10477360107681574608}{3617}p^{17}+\frac{11475149226459112710}{3617}p^{16}\\
&+\frac{10477360107681574608}{3617}p^{15}+\frac{7966990925710242000}{3617}p^{14}+\frac{5029685780950817520}{3617}p^{13}\\
&+\frac{2622064678835063496}{3617}p^{12}+\frac{1119620321590028304}{3617}p^{11}+\frac{387088285203107184}{3617}p^{10}\\
&+\frac{106630640734839600}{3617}p^{9}+\frac{22884510224189340}{3617}p^{8}+\frac{3705935707430928}{3617}p^{7}\\
&+\frac{431880430212432}{3617}p^{6}+\frac{33608470215088}{3617}p^{5}+\frac{1532766571768}{3617}p^{4}\\
&+\frac{31279681360}{3617}p^{3}+\frac{134229104}{3617}p^{2}+\frac{66032}{3617}p+1\Big)k^{16},
\end{aligned}
$$

$$
\begin{aligned}
E_{16}\left(q^{2}\right)=\Big(&p^{32}+16p^{31}+\frac{409184}{3617}p^{30}+\frac{1652680}{3617}p^{29}+\frac{37454668}{3617}p^{28}\\
&+\frac{473232088}{3617}p^{27}+\frac{6483541032}{3617}p^{26}+\frac{56895857928}{3617}p^{25}\\
&+\frac{349784703630}{3617}p^{24}+\frac{1625990384280}{3617}p^{23}+\frac{5904566270904}{3617}p^{22}\\
&+\frac{170854449814024}{3617}p^{21}\frac{400012993999716}{3617}p^{20}+\frac{76746190229880}{3617}p^{19}\\
&+\frac{121562792943720}{3617}p^{18}+\frac{159871907133288}{3617}p^{17}+\frac{175100498074755}{3617}p^{16}\\
&+\frac{159871907133288}{3617}p^{15}+\frac{121562792943720}{3617}p^{14}+\frac{76746190229880}{3617}p^{13}\\
&+\frac{400012993999716}{3617}p^{12}\frac{170854449814024}{3617}p^{11}+\frac{5904566270904}{3617}p^{10}\\
&+\frac{1625990384280}{3617}p^{9}+\frac{349784703630}{3617}p^{8}+\frac{56895857928}{3617}p^{7}\\
&+\frac{6483541032}{3617}p^{6}+\frac{473232088}{3617}p^{5}+\frac{37454668}{3617}p^{4}+\frac{1652680}{3617}p^{3}\\
&+\frac{409184}{3617}p^{2}+16p+1\Big)k^{16},
\end{aligned}
$$

$$E_{16}\left(q^3\right)=\left(p^{32}+16p^{31}+112p^{30}+\frac{1593520}{3617}p^{29}+\frac{3661048}{3617}p^{28}+\frac{4034128}{3617}p^{27}\right.$$
$$+\frac{6253392}{3617}p^{26}+\frac{94495728}{3617}p^{25}+\frac{584550300}{3617}p^{24}+\frac{2266437840}{3617}p^{23}$$
$$+\frac{8171962224}{3617}p^{22}+\frac{26676060144}{3617}p^{21}+\frac{64864877256}{3617}p^{20}$$
$$+\frac{116064498960}{3617}p^{19}+\frac{175509366480}{3617}p^{18}+\frac{243644509488}{3617}p^{17}$$
$$+\frac{279458260230}{3617}p^{16}+\frac{243644509488}{3617}p^{15}+\frac{175509366480}{3617}p^{14}$$
$$+\frac{116064498960}{3617}p^{13}+\frac{64864877256}{3617}p^{12}+\frac{26676060144}{3617}p^{11}$$
$$+\frac{8171962224}{3617}p^{10}+\frac{2266437840}{3617}p^{9}+\frac{584550300}{3617}p^{8}+\frac{94495728}{3617}p^{7}$$
$$\left.+\frac{6253392}{3617}p^{6}+\frac{4034128}{3617}p^{5}+\frac{3661048}{3617}p^{4}+\frac{1593520}{3617}p^{3}+112p^{2}+16p+1\right)k^{16}$$

$$E_{16}\left(q^4\right)=\left(\frac{1}{65536}p^{32}+\frac{3107}{14815232}p^{31}+\frac{8373509}{14815232}p^{30}-\frac{426024575}{3703808}p^{29}+\frac{21371210189}{7407616}p^{28}\right.$$
$$-\frac{2780210933}{231488}p^{27}-\frac{26515040733}{925952}p^{26}+\frac{46668005109}{462976}p^{25}+\frac{318947423355}{1851904}p^{24}$$
$$-\frac{132759215175}{462976}p^{23}-\frac{243776400843}{462976}p^{22}+\frac{49438651083}{115744}p^{21}+\frac{265269995439}{231488}p^{20}$$
$$+\frac{31565989785}{115744}p^{19}-\frac{6536535195}{14468}p^{18}+\frac{20783548953}{57872}p^{17}+\frac{573453480645}{462976}p^{16}$$
$$+\frac{58547998863}{57872}p^{15}+\frac{24046100565}{57872}p^{14}+\frac{2526073485}{14468}p^{13}+\frac{4307305053}{28936}p^{12}$$
$$+\frac{751890993}{7234}p^{11}+\frac{152474619}{3617}p^{10}+\frac{22732365}{3617}p^{9}-\frac{69921135}{14468}p^{8}-\frac{15090777}{3617}p^{7}$$
$$\left.-\frac{2659113}{3617}p^{6}+\frac{3862768}{3617}p^{5}+\frac{3632488}{3617}p^{4}+440p^{3}+112p^{2}+16p+1\right)k^{16}$$

$$E_{16}\left(q^6\right)=\left(p^{32}+16p^{31}+112p^{30}+440p^{29}+1004p^{28}+1064p^{27}-\frac{2745048}{3617}p^{26}\right.$$
$$-\frac{15372552}{3617}p^{25}-\frac{20069490}{3617}p^{24}-\frac{2603160}{3617}p^{23}+\frac{25848504}{3617}p^{22}$$
$$+\frac{33201144}{3617}p^{21}+\frac{8374116}{3617}p^{20}-\frac{19626360}{3617}p^{19}-\frac{18930840}{3617}p^{18}$$
$$+\frac{4834968}{3617}p^{17}+\frac{18832515}{3617}p^{16}+\frac{4834968}{3617}p^{15}-\frac{18930840}{3617}p^{14}$$
$$-\frac{19626360}{3617}p^{13}+\frac{8374116}{3617}p^{12}+\frac{33201144}{3617}p^{11}+\frac{25848504}{3617}p^{10}$$
$$-\frac{2603160}{3617}p^{9}-\frac{20069490}{3617}p^{8}-\frac{15372552}{3617}p^{7}-\frac{2745048}{3617}p^{6}$$
$$\left.+1064p^{5}+1004p^{4}+440p^{3}+112p^{2}+16p+1\right)k^{16}$$

$$E_{16}\left(q^{12}\right)=\left(\frac{1}{65536}p^{32}+\frac{1}{4096}p^{31}+\frac{7}{4096}p^{30}+\frac{24835}{3703808}p^{29}\right.$$
$$+\frac{112559}{7407616}p^{28}+\frac{3577}{231488}p^{27}+\frac{19407}{925952}p^{26}+\frac{179709}{462976}p^{25}$$
$$+\frac{4552575}{1851904}p^{24}+\frac{2646075}{462976}p^{23}-\frac{4566453}{462976}p^{22}-\frac{11870997}{115744}p^{21}$$
$$-\frac{64064691}{231488}p^{20}-\frac{24289965}{115744}p^{19}+\frac{40724415}{57872}p^{18}$$
$$+\frac{127224993}{57872}p^{17}+\frac{932409285}{462976}p^{16}-\frac{109346667}{57872}p^{15}$$
$$-\frac{386476065}{57872}p^{14}-\frac{82555605}{14468}p^{13}+\frac{67084983}{28936}p^{12}$$
$$+\frac{18339}{2}p^{11}+7122p^{10}-735p^{9}-\frac{22215}{4}p^{8}-4251p^{7}$$
$$\left.-759p^{6}+1064p^{5}+1004p^{4}+440p^{3}+112p^{2}+16p+1\right)k^{16}.$$

It is easy to check the following expressions by (20)-(25)

$$f_1:=\sum_{n=0}^{\infty}f_1(n)=\frac{\eta^{18}(4z)\eta^{16}(6z)\eta^{10}(12z)}{\eta^{12}(2z)}$$
$$=\left(-\frac{1}{524288}p^{27}-\frac{45}{1048576}p^{26}-\frac{29}{65536}p^{25}\right.$$
$$-\frac{2897}{1048576}p^{24}-\frac{761}{65536}p^{23}-\frac{36143}{1048576}p^{22}$$
$$-\frac{4809}{65536}p^{21}-\frac{115427}{1048576}p^{20}-\frac{55863}{524288}p^{19}$$
$$-\frac{2653}{65536}p^{18}+\frac{1845}{32768}p^{17}+\frac{3859}{32768}p^{16}+\frac{1823}{16384}p^{15}$$
$$\left.+\frac{133}{2048}p^{14}+\frac{49}{2048}p^{13}+\frac{21}{4096}p^{12}+\frac{1}{2048}p^{11}\right)k^{16},$$

$$f_2=\sum_{n=0}^{\infty}f_2(n)=\frac{\eta^{20}(4z)\eta^{2}(6z)\eta^{20}(12z)}{\eta^{10}(2z)}$$
$$=\left(-\frac{1}{8388608}p^{29}-\frac{49}{16777216}p^{28}-\frac{69}{2097152}p^{27}\right.$$
$$-\frac{945}{4194304}p^{26}\frac{4377}{4194304}p^{25}-\frac{57639}{16777216}p^{2422}$$
$$-\frac{68717}{8388608}p^{23}-\frac{58817}{4194304}p-\frac{34155}{2097152}p^{21}$$
$$-\frac{5291}{524288}p^{20}+\frac{847}{262144}p^{19}+\frac{1959}{131072}p^{18}+\frac{1137}{65536}p^{17}$$
$$\left.+\frac{765}{65536}p^{16}+\frac{159}{32768}p^{15}+\frac{19}{16384}p^{14}+\frac{1}{8192}p^{13}\right)k^{16},$$

$$f_3=\sum_{n=0}^{\infty}f_3(n)=\frac{\eta^{15}(4z)\eta^{7}(6z)\eta^{19}(12z)}{\eta^{9}(2z)}$$
$$=\left(-\frac{1}{2097152}p^{28}-\frac{43}{4194304}p^{27}-\frac{421}{4194304}p^{26}-\frac{619}{1048576}p^{25}-\frac{2425}{1048576}p^{24}\right.$$
$$-\frac{26443}{4194304}p^{23}-\frac{50501}{4194304}p^{22}-\frac{32463}{2097152}p^{21}-\frac{2925}{262144}p^{20}+\frac{17}{16384}p^{19}$$
$$\left.+\frac{1709}{131072}p^{18}+\frac{1075}{65536}p^{17}+\frac{187}{16384}p^{16}+\frac{79}{16384}p^{15}+\frac{19}{16384}p^{14}+\frac{1}{8192}p^{13}\right)k^{16},$$

$$f_4=\sum_{n=0}^{\infty}f_4(n)=\frac{\eta^{18}(12z)\eta^{10}(4z)\eta^{12}(6z)}{\eta^{8}(2z)}=\left(-\frac{1}{524288}p^{27}-\frac{37}{1048576}p^{26}\right.$$
$$-\frac{77}{262144}p^{25}-\frac{1517}{1048576}p^{24}-\frac{1219}{262144}p^{23}-\frac{10571}{1048576}p^{22}-\frac{3789}{262144}p^{21}$$
$$-\frac{12519}{1048576}p^{20}-\frac{513}{524288}p^{19}+\frac{2933}{262144}p^{18}+\frac{2027}{131072}p^{17}$$
$$\left.+\frac{731}{65536}p^{16}+\frac{157}{32768}p^{15}+\frac{19}{16384}p^{14}+\frac{1}{8192}p^{13}\right)k^{16},$$

$$f_5=\sum_{n=0}^{\infty}f_5(n)=\frac{\eta^{5}(4z)\eta^{17}(6z)\eta^{17}(12z)}{\eta^{7}(2z)}=\left(-\frac{1}{131072}p^{26}-\frac{31}{262144}p^{25}-\frac{213}{262144}p^{24}\right.$$
$$-\frac{849}{262144}p^{23}-\frac{2145}{262144}p^{22}-\frac{3471}{262144}p^{21}-\frac{3263}{262144}p^{20}-\frac{739}{262144}p^{19}+\frac{2463}{262144}p^{18}$$
$$\left.+\frac{1905}{131072}p^{17}+\frac{357}{32768}p^{16}+\frac{39}{8192}p^{15}+\frac{19}{16384}p^{14}+\frac{1}{8192}p^{13}\right)k^{16},$$

$$f_6=\sum_{n=0}^{\infty}f_6(n)=\frac{\eta^{17}(4z)\eta^{17}(6z)\eta^{5}(12z)}{\eta^{7}(2z)}$$

$$=\left(\frac{1}{65536}p^{27}+\frac{11}{32768}p^{26}+\frac{881}{262144}p^{25}+\frac{5285}{262144}p^{24}+\frac{20991}{262144}p^{23}\right.$$
$$+\frac{57213}{262144}p^{22}+\frac{105569}{262144}p^{21}+\frac{117767}{262144}p^{20}+\frac{31081}{262144}p^{19}-\frac{142257}{262144}p^{18}$$
$$-\frac{136127}{131072}p^{17}-\frac{14399}{16384}p^{16}-\frac{397}{2048}p^{15}+\frac{3249}{8192}p^{14}$$
$$\left.+\frac{2103}{4096}p^{13}+\frac{161}{512}p^{12}+\frac{115}{1024}p^{11}+\frac{23}{1024}p^{10}+\frac{1}{512}p^{9}\right)k^{16},$$

$$f_7=\sum_{n=0}^{\infty}f_7(n)=\frac{\eta^{19}(4z)\eta^{15}(6z)\eta^{15}(12z)}{\eta^{17}(2z)}$$

$$=\left(\frac{1}{4194304}p^{27}+\frac{23}{4194304}p^{26}+\frac{61}{1048576}p^{25}+\frac{791}{2097152}p^{24}+\frac{7001}{4194304}p^{23}\right.$$
$$+\frac{22363}{4194304}p^{22}+\frac{26577}{2097152}p^{21}+\frac{5967}{262144}p^{20}+\frac{1017}{32768}p^{19}+\frac{4189}{131072}p^{18}$$
$$\left.+\frac{1603}{65536}p^{17}+\frac{221}{16384}p^{16}+\frac{83}{16384}p^{15}+\frac{19}{16384}p^{14}+\frac{1}{8192}p^{13}\right)k^{16},$$

$$f_8=\sum_{n=0}^{\infty}f_8(n)=\frac{\eta^{17}(2z)\eta^{17}(4z)\eta^{5}(12z)}{\eta^{7}(6z)}$$

$$=\left(\frac{1}{4096}p^{31}+\frac{3}{512}p^{30}+\frac{1031}{16384}p^{29}+\frac{6445}{16384}p^{28}+\frac{101019}{65536}p^{27}\right.$$
$$+\frac{120001}{32768}p^{26}+\frac{984145}{262144}p^{25}-\frac{1537659}{262144}p^{24}-\frac{7102105}{262144}p^{23}$$
$$-\frac{9302339}{262144}p^{22}+\frac{2219505}{262144}p^{21}+\frac{23214863}{262144}p^{20}+\frac{26270329}{262144}p^{19}$$
$$-\frac{5377425}{262144}p^{18}-\frac{20003423}{131072}p^{17}-\frac{3918925}{32768}p^{16}+\frac{346941}{8192}p^{15}$$
$$+\frac{2088461}{16384}p^{14}+\frac{512615}{8192}p^{13}-\frac{7401}{256}p^{12}-\frac{47239}{1024}p^{11}$$
$$\left.-\frac{16583}{1024}p^{10}+\frac{2319}{512}p^{9}+\frac{775}{128}p^{8}+\frac{37}{16}p^{7}+\frac{27}{64}p^{6}+\frac{1}{32}p^{5}\right)k^{16},$$

$$f_9=\sum_{n=0}^{\infty}f_9(n)=\frac{\eta^{20}(2z)\eta^{14}(4z)\eta^{14}(12z)}{\eta^{16}(6z)}$$

$$=\left(\frac{1}{16384}p^{32}+\frac{23}{16384}p^{31}+\frac{939}{65536}p^{30}+\frac{2753}{32768}p^{29}+\frac{78995}{262144}p^{28}\right.$$

$$+\frac{161007}{262144}p^{27}+\frac{340117}{1048576}p^{26}-\frac{117361}{65536}p^{25}-\frac{5224329}{1048576}p^{24}$$

$$-\frac{2039005}{524288}p^{23}+\frac{6297515}{1048576}p^{22}+\frac{4229337}{262144}p^{21}+\frac{9352981}{1048576}p^{20}$$

$$-\frac{7365203}{524288}p^{19}-\frac{3159555}{131072}p^{18}-\frac{379685}{65536}p^{17}+\frac{1073567}{65536}p^{16}$$

$$+\frac{507447}{32768}p^{15}+\frac{323}{2048}p^{14}-\frac{15125}{2048}p^{13}-\frac{16989}{4096}p^{12}$$

$$\left.+\frac{203}{2048}p^{11}+\frac{529}{512}p^{10}+\frac{123}{256}p^{9}+\frac{25}{256}p^{8}+\frac{1}{128}p^{7}\right)k^{16},$$

$$f_{10}=\sum_{n=0}^{\infty}f_{10}(n)=\frac{\eta^{16}(4z)\eta^{18}(6z)}{\eta^{2}(2z)}$$

$$=\left(-\frac{1}{8192}p^{27}-\frac{43}{16384}p^{26}-\frac{835}{32768}p^{25}-\frac{9601}{65536}p^{24}\right.$$

$$-\frac{4477}{8192}p^{23}-\frac{44075}{32768}p^{22}-\frac{66467}{32768}p^{21}-\frac{4547}{4096}p^{20}$$

$$+\frac{80587}{32768}p^{19}+\frac{216681}{32768}p^{18}+\frac{108333}{16384}p^{17}+\frac{46257}{65536}p^{16}$$

$$-\frac{200503}{32768}p^{15}-\frac{30571}{4096}p^{14}-\frac{6751}{2048}p^{13}+\frac{2303}{2048}p^{12}$$

$$\left.+\frac{2471}{1024}p^{11}+\frac{391}{256}p^{10}+\frac{67}{128}p^{9}+\frac{25}{256}p^{8}+\frac{1}{128}p^{7}\right)k^{16},$$

$$f_{11}=\sum_{n=0}^{\infty}f_{11}(n)=\frac{\eta^{20}(2z)\eta^{14}(4z)\eta^{8}(6z)}{\eta^{10}(12z)}$$

$$=\left(-\frac{1}{32}p^{29}-\frac{43}{64}p^{28}-\frac{813}{128}p^{27}-\frac{8683}{256}p^{26}-\frac{54571}{512}p^{25}\right.$$

$$-\frac{169029}{1024}p^{24}+\frac{166177}{2048}p^{23}+\frac{3773287}{4096}p^{22}+\frac{1665717}{1024}p^{21}$$

$$+\frac{541157}{4096}p^{20}-\frac{8039515}{2048}p^{19}-\frac{11541531}{2048}p^{18}+\frac{52297}{128}p^{17}$$

$$+\frac{19076567}{2048}p^{16}+\frac{17486919}{2048}p^{15}-\frac{13136125}{4096}p^{14}$$

$$-\frac{11342417}{1024}p^{13}-\frac{22863279}{4096}p^{12}+\frac{8459683}{2048}p^{11}$$

$$+\frac{6273043}{1024}p^{10}+\frac{828501}{512}p^{9}-\frac{220715}{128}p^{8}-\frac{98825}{64}p^{7}$$

$$\left.-\frac{10071}{32}p^{6}+\frac{3281}{16}p^{5}+\frac{2581}{16}p^{4}+\frac{399}{8}p^{3}+\frac{31}{4}p^{2}+\frac{1}{2}p\right)k^{16},$$

$$f_{12}=\sum_{n=0}^{\infty}f_{12}(n)=\eta^{20}(2z)\eta^{2}(4z)\eta^{8}(6z)\eta^{2}(12z)$$

$$=\left(\frac{1}{64}p^{28}+\frac{15}{64}p^{27}+\frac{371}{256}p^{26}+\frac{569}{128}p^{25}+\frac{5211}{1024}p^{24}-\frac{9241}{1024}p^{23}\right.$$

$$-\frac{153891}{4096}p^{22}-\frac{8919}{256}p^{21}+\frac{179109}{4096}p^{20}+\frac{251003}{2048}p^{19}+\frac{113727}{2048}p^{18}$$

$$-\frac{30079}{256}p^{17}-\frac{329953}{2048}p^{16}-\frac{2349}{256}p^{15}+\frac{497249}{4096}p^{14}+\frac{42867}{512}p^{13}$$

$$-\frac{63711}{4096}p^{12}-\frac{93897}{2048}p^{11}-\frac{19619}{1024}p^{10}+\frac{1587}{512}p^{9}+\frac{1475}{256}p^{8}$$

$$\left.+\frac{293}{128}p^{7}+\frac{27}{64}p^{6}+\frac{1}{32}p^{5}\right)k^{16},$$

$$f_{14}=\sum_{n=0}^{\infty}f_{14}(n)=\frac{\eta^{16}(2z)\eta^{10}(4z)\eta^{18}(12z)}{\eta^{12}(6z)}$$

$$=\left(-\frac{1}{32768}p^{31}-\frac{41}{65536}p^{30}-\frac{369}{65536}p^{29}-\frac{3755}{131072}p^{28}\right.$$
$$-\frac{45365}{524288}p^{27}-\frac{142317}{1048576}p^{26}+\frac{2541}{262144}p^{25}+\frac{523731}{1048576}p^{24}$$
$$+\frac{238005}{262144}p^{23}+\frac{289525}{1048576}p^{22}-\frac{358967}{262144}p^{21}-\frac{2160639}{1048576}p^{20}$$
$$-\frac{178493}{524288}p^{19}+\frac{503675}{262144}p^{18}+\frac{246645}{131072}p^{17}-\frac{51}{32768}p^{16}$$
$$-\frac{18129}{16384}p^{15}-\frac{5643}{8192}p^{14}+\frac{25}{4096}p^{13}+\frac{805}{4096}p^{12}+\frac{207}{2048}p^{11}$$
$$\left.+\frac{23}{1024}p^{10}+\frac{1}{512}p^{9}\right)k^{16},$$

$$f_{15}=\sum_{n=0}^{\infty}f_{15}(n)=\frac{\eta^{19}(4z)\eta^{9}(6z)\eta^{15}(12z)}{\eta^{11}(2z)}$$

$$=\left(-\frac{1}{2097152}p^{28}-\frac{47}{4194304}p^{27}-\frac{507}{4194304}p^{26}-\frac{1659}{2097152}p^{25}\right.$$
$$-\frac{3663}{1048576}p^{24}-\frac{45843}{4194304}p^{23}-\frac{103387}{4194304}p^{22}-\frac{20741}{524288}p^{21}$$
$$-\frac{44163}{1048576}p^{20}-\frac{2789}{131072}p^{19}+\frac{1981}{131072}p^{18}+\frac{87}{2048}p^{17}$$
$$\left.+\frac{1449}{32768}p^{16}+\frac{453}{16384}p^{15}+\frac{177}{16384}p^{14}+\frac{5}{2048}p^{13}+\frac{1}{4096}p^{12}\right)k^{16},$$

$$f_{16}=\sum_{n=0}^{\infty}f_{16}(n)=\frac{\eta^{14}(4z)\eta^{14}(6z)\eta^{14}(12z)}{\eta^{10}(2z)}$$

$$=\left(-\frac{1}{524288}p^{27}-\frac{41}{1048576}p^{26}-\frac{191}{524288}p^{25}-\frac{2133}{1048576}p^{24}\right.$$
$$-\frac{3955}{524288}p^{23}-\frac{20323}{1048576}p^{22}-\frac{18149}{524288}p^{21}-\frac{42831}{1048576}p^{20}$$
$$-\frac{1629}{65536}p^{19}+\frac{605}{65536}p^{18}+\frac{155}{4096}p^{17}+\frac{1379}{32768}p^{16}$$
$$\left.+\frac{111}{4096}p^{15}+\frac{11}{1024}p^{14}+\frac{5}{2048}p^{13}+\frac{1}{4096}p^{12}\right)k^{16},$$

$$f_{17}=\sum_{n=0}^{\infty}f_{17}(n)=\frac{\eta^{15}(4z)\eta^{13}(6z)\eta^{19}(12z)}{\eta^{15}(2z)}$$

$$=\left(\frac{1}{4194304}p^{27}+\frac{21}{4194304}p^{26}+\frac{101}{2097152}p^{25}+\frac{589}{2097152}p^{24}\right.$$

$$+\frac{4645}{4194304}p^{23}+\frac{13073}{4194304}p^{22}+\frac{211}{32768}p^{21}+\frac{2591}{262144}p^{20}$$

$$+\frac{1477}{131072}p^{19}+\frac{1235}{131072}p^{18}+\frac{23}{4096}p^{17}$$

$$\left.+\frac{37}{16384}p^{16}+\frac{9}{16384}p^{15}+\frac{1}{16384}p^{14}\right)k^{16},$$

$$f_{18}=\sum_{n=0}^{\infty}f_{18}(n)=\frac{\eta^{9}(4z)\eta^{19}(6z)\eta^{13}(12z)}{\eta^{9}(2z)}$$

$$=\left(-\frac{1}{131072}p^{26}-\frac{35}{262144}p^{25}-\frac{275}{262144}p^{24}-\frac{1275}{262144}p^{23}\right.$$

$$-\frac{3843}{262144}p^{22}-\frac{7761}{262144}p^{21}-\frac{10205}{262144}p^{20}-\frac{7265}{262144}p^{19}$$

$$+\frac{985}{262144}p^{18}+\frac{273}{8192}p^{17}+\frac{2619}{65536}p^{16}+\frac{435}{16384}p^{15}$$

$$\left.+\frac{175}{16384}p^{14}+\frac{5}{2048}p^{13}+\frac{1}{4096}p^{12}\right)k^{16},$$

$$f_{19}=\sum_{n=0}^{\infty}f_{19}(n)=\frac{\eta^{18}(4z)\eta^{10}(6z)\eta^{10}(12z)}{\eta^{6}(2z)}$$

$$=\left(\frac{1}{262144}p^{28}+\frac{23}{262144}p^{27}+\frac{965}{1048576}p^{26}+\frac{3041}{524288}p^{25}\right.$$

$$+\frac{25479}{1048576}p^{24}+\frac{18429}{262144}p^{23}+\frac{146279}{1048576}p^{22}+\frac{91313}{524288}p^{21}$$

$$+\frac{83989}{1048576}p^{20}-\frac{41021}{262144}p^{19}-\frac{98171}{262144}p^{18}-\frac{1493}{4096}p^{17}$$

$$-\frac{4043}{32768}p^{16}+\frac{1029}{8192}p^{15}+\frac{1647}{8192}p^{14}+\frac{275}{2048}p^{13}$$

$$\left.+\frac{209}{4096}p^{12}+\frac{11}{1024}p^{11}+\frac{1}{1024}p^{10}\right)k^{16},$$

$$f_{20}=\sum_{n=0}^{\infty}f_{20}(n)=\frac{\eta^{20}(2z)\eta^{8}(4z)\eta^{20}(6z)}{\eta^{16}(12z)}$$

$$=\left(-\frac{1}{2}p^{27}-\frac{35}{4}p^{26}-\frac{521}{8}p^{25}-\frac{4111}{16}p^{24}-\frac{15975}{32}p^{23}+\frac{327}{64}p^{22}\right.$$

$$+\frac{297757}{128}p^{21}+\frac{1195499}{256}p^{20}+\frac{106085}{128}p^{19}-\frac{685735}{64}p^{18}$$

$$-\frac{62285}{4}p^{17}+\frac{435395}{256}p^{16}+\frac{1648595}{64}p^{15}+\frac{1378995}{64}p^{14}$$

$$-\frac{1378995}{128}p^{13}-\frac{7434175}{256}p^{12}-\frac{1547495}{128}p^{11}+\frac{776305}{64}p^{10}$$

$$+\frac{951275}{64}p^{9}+\frac{734305}{256}p^{8}-\frac{73567}{16}p^{7}-\frac{56545}{16}p^{6}-\frac{2247}{4}p^{5}$$

$$\left.+\frac{4203}{8}p^{4}+365p^{3}+107p^{2}+16p+1\right)k^{16},$$

$$f_{21}=\sum_{n=0}^{\infty}f_{21}(n)=\frac{\eta^{20}(2z)\eta^{20}(4z)}{\eta^{4}(6z)\eta^{4}(12z)}$$

$$=\left(-\frac{1}{512}p^{31}-\frac{51}{1024}p^{30}-\frac{1169}{2048}p^{29}-\frac{15687}{4096}p^{28}-\frac{133119}{8192}p^{27}\right.$$

$$-\frac{697473}{16384}p^{26}-\frac{1714923}{32768}p^{25}+\frac{3811059}{65536}p^{24}+\frac{11844819}{32768}p^{23}$$

$$+\frac{18633153}{32768}p^{22}-\frac{456813}{32768}p^{21}-\frac{23799867}{16384}p^{20}-\frac{67065705}{32768}p^{19}$$

$$+\frac{3013731}{32768}p^{18}+\frac{27673335}{8192}p^{17}+\frac{219131721}{65536}p^{16}-\frac{14731353}{16384}p^{15}$$

$$-\frac{68551245}{16384}p^{14}-\frac{5076561}{2048}p^{13}+\frac{5524785}{4096}p^{12}+\frac{2546313}{1024}p^{11}$$

$$+\frac{856899}{1024}p^{10}-\frac{40113}{64}p^{9}-\frac{170541}{256}p^{8}-\frac{10647}{64}p^{7}$$

$$\left.+\frac{4977}{64}p^{6}+\frac{567}{8}p^{5}+\frac{371}{16}p^{4}+\frac{15}{4}p^{3}+\frac{1}{4}p^{2}\right)k^{16},$$

$$
\begin{aligned}
f_{22} &= \sum_{n=0}^{\infty} f_{22}(n) = \frac{\eta^{20}(4z)\eta^{20}(12z)}{\eta^{4}(2z)\eta^{4}(6z)} \\
&= \left(\frac{1}{4194304} p^{30} + \frac{25}{4194304} p^{29} + \frac{1145}{16777216} p^{28} \right. \\
&\quad + \frac{495}{1048576} p^{27} + \frac{18341}{8388608} p^{26} + \frac{29669}{4194304} p^{25} \\
&\quad + \frac{269025}{16777216} p^{24} + \frac{50735}{2097152} p^{23} + \frac{39853}{2097152} p^{22} \\
&\quad - \frac{4411}{524288} p^{21} - \frac{46849}{1048576} p^{20} - \frac{3785}{65536} p^{19} \\
&\quad - \frac{2115}{65536} p^{18} + \frac{67}{8192} p^{17} + \frac{1903}{65536} p^{16} + \frac{195}{8192} p^{15} \\
&\quad \left. + \frac{85}{8192} p^{14} + \frac{5}{2048} p^{13} + \frac{1}{4096} p^{12} \right) k^{16},
\end{aligned}
$$

$$
\begin{aligned}
f_{23} &= \sum_{n=0}^{\infty} f_{23}(n) = \frac{\eta^{20}(4z)\eta^{20}(6z)}{\eta^{4}(2z)\eta^{4}(12z)} \\
&= \left(-\frac{1}{8192} p^{27} - \frac{47}{16384} p^{26} - \frac{1007}{32768} p^{25} - \frac{12941}{65536} p^{24} \right. \\
&\quad - \frac{27509}{32768} p^{23} - \frac{79891}{32768} p^{22} - \frac{154617}{32768} p^{21} - \frac{84655}{16384} p^{20} \\
&\quad + \frac{7835}{32768} p^{19} + \frac{377855}{32768} p^{18} + \frac{162507}{8192} p^{17} + \frac{912921}{65536} p^{16} \\
&\quad - \frac{77123}{16384} p^{15} - \frac{322787}{16384} p^{14} - \frac{18661}{1024} p^{13} - \frac{11199}{2048} p^{12} \\
&\quad + \frac{2387}{512} p^{11} + \frac{3253}{512} p^{10} + \frac{229}{64} p^{9} + \frac{293}{256} p^{8} \\
&\quad \left. + \frac{13}{64} p^{7} + \frac{1}{64} p^{6} \right) k^{16},
\end{aligned}
$$

$$f_{24}=\sum_{n=0}^{\infty}f_{24}(n)=\frac{\eta^{15}(4z)\eta(6z)\eta^{19}(12z)}{\eta^{3}(2z)}$$
$$=\left(\frac{1}{1048576}p^{29}+\frac{11}{524288}p^{28}+\frac{877}{4194304}p^{27}+\frac{5205}{4194304}p^{26}\right.$$
$$+\frac{10137}{2097152}p^{25}+\frac{26721}{2097152}p^{24}+\frac{92837}{4194304}p^{23}+\frac{89789}{4194304}p^{22}$$
$$-\frac{725}{524288}p^{21}-\frac{39571}{1048576}p^{20}-\frac{7325}{131072}p^{19}-\frac{4619}{131072}p^{18}$$
$$+\frac{9}{2048}p^{17}+\frac{885}{32768}p^{16}+\frac{381}{16384}p^{15}+\frac{169}{16384}p^{14}$$
$$\left.+\frac{5}{2048}p^{13}+\frac{1}{4096}p^{12}\right)k^{16},$$

$$f_{25}=\sum_{n=0}^{\infty}f_{25}(n)=\frac{\eta^{20}(2z)\eta^{20}(4z)}{\eta^{4}(6z)\eta^{4}(12z)}$$
$$=\left(-\frac{1}{8}p^{25}-\frac{9}{16}p^{24}-\frac{7}{16}p^{23}+\frac{55}{32}p^{22}+\frac{435}{128}p^{21}\right.$$
$$-\frac{101}{256}p^{20}-\frac{781}{128}p^{19}-\frac{33}{8}p^{18}+\frac{115}{32}p^{17}+\frac{1475}{256}p^{16}$$
$$+\frac{63}{64}p^{15}-\frac{41}{16}p^{14}-\frac{229}{128}p^{13}-\frac{15}{256}p^{12}$$
$$\left.+\frac{55}{128}p^{11}+\frac{7}{32}p^{10}+\frac{3}{64}p^{9}+\frac{1}{256}p^{8}\right)k^{16},$$

$$f_{26}=\sum_{n=0}^{\infty} f_{26}(n)=\frac{\eta^{20}(2z)\eta^{20}(6z)\eta^{8}(12z)}{\eta^{4}(4z)}$$

$$=\left(-\frac{1}{8}p^{28}-\frac{41}{16}p^{27}-\frac{735}{32}p^{26}-\frac{7369}{64}p^{25}-\frac{42461}{128}p^{24}\right.$$
$$-\frac{108327}{256}p^{23}+\frac{261427}{512}p^{22}+\frac{3059933}{1024}p^{21}$$
$$+\frac{4316289}{1024}p^{20}-\frac{1428553}{1024}p^{19}-\frac{12680767}{1024}p^{18}$$
$$-\frac{6900279}{512}p^{17}+\frac{3095779}{512}p^{16}+\frac{13721761}{512}p^{15}$$
$$+\frac{2279991}{128}p^{14}-\frac{14642143}{1024}p^{13}-\frac{29221507}{1024}p^{12}$$
$$-\frac{9789933}{1024}p^{11}+\frac{13635953}{1024}p^{10}+\frac{921227}{64}p^{9}$$
$$+\frac{560229}{256}p^{8}-\frac{153647}{32}p^{7}-\frac{111071}{32}p^{6}$$
$$\left.-\frac{987}{2}p^{5}+\frac{4387}{8}p^{4}+\frac{1475}{4}p^{3}+\frac{429}{4}p^{2}+16p+1\right)k^{16},$$

$$f_{27}=\sum_{n=0}^{\infty} f_{27}(n)=\frac{\eta^{19}(2z)\eta^{13}(4z)\eta^{15}(6z)}{\eta^{15}(12z)}$$

$$=\left(-\frac{1}{32}p^{26}-\frac{15}{64}p^{25}-\frac{9}{16}p^{24}-\frac{1}{128}p^{23}\right.$$
$$+\frac{1095}{512}p^{22}+\frac{2949}{1024}p^{21}-\frac{1435}{1024}p^{20}$$
$$-\frac{6273}{1024}p^{19}-\frac{3279}{1024}p^{18}+\frac{2105}{512}p^{17}+\frac{2811}{512}p^{16}$$
$$+\frac{315}{512}p^{15}-\frac{169}{64}p^{14}-\frac{1767}{1024}p^{13}-\frac{15}{1024}p^{12}$$
$$\left.+\frac{451}{1024}p^{11}+\frac{225}{1024}p^{10}+\frac{3}{64}p^{9}+\frac{1}{256}p^{8}\right)k^{16}.$$

Obviously, $f_1,\cdots,f_{27}$ are functions of q, see (3), (19). We see that

$$f_7,f_9,f_{17},f_{20},f_{25},f_{26},f_{27}\in M_{16}\left(\Gamma_0(12)\right)\setminus S_{16}\left(\Gamma_0(12)\right)$$

by [17] . Now

$$
\begin{aligned}
&\eta^{a_1}(z)\eta^{a_2}(2z)\eta^{a_3}(3z)\eta^{a_4}(4z)\eta^{a_6}(6z)\eta^{a_{12}}(12z)\\
&=q^{b_1}\prod_{n=1}^{\infty}\left(1-q^{n}\right)^{a_1}\left(1-q^{2n}\right)^{a_2}\left(1-q^{3n}\right)^{a_3}\left(1-q^{4n}\right)^{a_4}\left(1-q^{6n}\right)^{a_6}\left(1-q^{12n}\right)^{a_{12}}\\
&=2^{-\frac{a_1}{6}-\frac{a_2}{3}-\frac{a_3}{6}-\frac{2a_4}{3}-\frac{a_6}{3}-\frac{2a_{12}}{3}}\,p^{\frac{a_1}{24}+\frac{a_2}{12}+\frac{a_3}{8}+\frac{a_4}{6}+\frac{a_6}{4}+\frac{a_{12}}{2}}\\
&\quad\cdot(1-p)^{\frac{a_1}{2}+\frac{a_2}{4}+\frac{a_3}{6}+\frac{a_4}{8}+\frac{a_6}{12}+\frac{a_{12}}{24}}(1+p)^{\frac{a_1}{6}+\frac{a_2}{12}+\frac{a_3}{2}+\frac{a_4}{24}+\frac{a_6}{4}+\frac{a_{12}}{8}}\\
&\quad\cdot(1+2p)^{\frac{a_1}{8}+\frac{a_2}{4}+\frac{a_3}{24}+\frac{a_4}{8}+\frac{a_6}{12}+\frac{a_{12}}{24}}(2+p)^{\frac{a_1}{8}+\frac{a_2}{4}+\frac{a_3}{24}+\frac{a_4}{2}+\frac{a_6}{12}+\frac{a_{12}}{6}}\\
&\quad\cdot k^{\frac{a_1+a_2+a_3+a_4+a_6+a_{12}}{2}}\\
&=\frac{k^{16}}{2^{b_1+b_5}}p^{b_1}(1-p)^{b_2}(1+p)^{b_3}(1+2p)^{b_4}(2+p)^{b_5}\\
&=k^{16}\left(k_0+k_1p+k_2p^2+k_3p^3+k_4p^4+k_5p^5\right.\\
&\quad+k_6p^6+k_7p^7+k_8p^8+k_9p^9+k_{10}p^{10}+k_{11}p^{11}\\
&\quad+k_{12}p^{12}+k_{13}p^{13}+k_{14}p^{14}+k_{15}p^{15}+k_{16}p^{16}\\
&\quad+k_{17}p^{17}+k_{18}p^{18}+k_{19}p^{19}+k_{20}p^{20}+k_{21}p^{21}\\
&\quad+k_{22}p^{22}+k_{23}p^{23}+k_{24}p^{24}+k_{25}p^{25}+k_{26}p^{26}\\
&\quad\left.+k_{27}p^{27}+k_{28}p^{28}+k_{29}p^{29}+k_{30}p^{30}+k_{31}p^{31}+k_{32}p^{32}\right)\\
&=\frac{3617c_1}{16320}\left(1+\frac{16320}{3617}\sum_{n=1}^{\infty}\sigma_{15}(n)q^{n}\right)+\frac{3617c_2}{16320}\left(1+\frac{16320}{3617}\sum_{n=1}^{\infty}\sigma_{15}(n)q^{2n}\right)\\
&\quad+\frac{3617c_3}{16320}\left(1+\frac{16320}{3617}\sum_{n=1}^{\infty}\sigma_{15}(n)q^{3n}\right)+\frac{3617c_4}{16320}\left(1+\frac{16320}{3617}\sum_{n=1}^{\infty}\sigma_{15}(n)q^{4n}\right)\\
&\quad+\frac{3617c_5}{16320}\left(1+\frac{16320}{3617}\sum_{n=1}^{\infty}\sigma_{15}(n)q^{6n}\right)+\frac{3617c_6}{16320}\left(1+\frac{16320}{3617}\sum_{n=1}^{\infty}\sigma_{15}(n)q^{12n}\right)\\
&\quad+r_1q^{11}\prod_{n=1}^{\infty}\frac{\left(1-q^{4n}\right)^{18}\left(1-q^{6n}\right)^{16}\left(1-q^{12n}\right)^{10}}{\left(1-q^{2n}\right)^{12}}+r_2q^{13}\prod_{n=1}^{\infty}\frac{\left(1-q^{4n}\right)^{20}\left(1-q^{6n}\right)^{2}\left(1-q^{12}\right)^{20}}{\left(1-q^{2n}\right)^{10}}\\
&\quad+r_3q^{13}\prod_{n=1}^{\infty}\frac{\left(1-q^{4n}\right)^{15}\left(1-q^{6n}\right)^{7}\left(1-q^{12n}\right)^{19}}{\left(1-q^{2n}\right)^{9}}+r_4q^{13}\prod_{n=1}^{\infty}\frac{\left(1-q^{2n}\right)^{10}\left(1-q^{4n}\right)^{12}\left(1-q^{6n}\right)^{18}}{\left(1-q^{12n}\right)^{8}}\\
&\quad+r_5q^{13}\prod_{n=1}^{\infty}\frac{\left(\left(1-q^{4n}\right)^{7}(1-q^{6n}\right)^{17}\left(1-q^{12n}\right)^{17}}{\left(1-q^{2n}\right)^{7}}+r_6q^{9}\prod_{n=1}^{\infty}\frac{\left(1-q^{4n}\right)^{17}\left(1-q^{6n}\right)^{17}\left(1-q^{12n}\right)^{5}}{\left(1-q^{2n}\right)^{7}}\\
&\quad+r_7q^{13}\prod_{n=1}^{\infty}\frac{\left(1-q^{4n}\right)^{19}\left(1-q^{6n}\right)^{15}\left(1-q^{12n}\right)^{15}}{\left(1-q^{2n}\right)^{17}}+r_8q^{5}\prod_{n=1}^{\infty}\frac{\left(1-q^{2n}\right)^{17}\left(1-q^{4n}\right)^{17}\left(1-q^{12n}\right)^{5}}{\left(1-q^{6n}\right)^{7}}\\
&\quad+r_9q^{7}\prod_{n=1}^{\infty}\frac{\left(1-q^{2n}\right)^{20}\left(1-q^{4n}\right)^{14}\left(1-q^{12n}\right)^{14}}{\left(1-q^{6n}\right)^{16}}+r_{10}q^{7}\prod_{n=1}^{\infty}\frac{\left(1-q^{4n}\right)^{16}\left(1-q^{6n}\right)^{18}}{\left(1-q^{2n}\right)^{2}}\\
&\quad+r_{11}q\prod_{n=1}^{\infty}\frac{\left(1-q^{2n}\right)^{20}\left(1-q^{4n}\right)^{14}\left(1-q^{6n}\right)^{8}}{\left(1-q^{12n}\right)^{10}}+r_{12}q^{5}\prod_{n=1}^{\infty}\left(1-q^{2n}\right)^{20}\left(1-q^{4n}\right)^{2}\left(1-q^{6n}\right)^{8}\left(1-q^{12n}\right)^{2}
\end{aligned}
$$

$$+r_{13}q^{13}\prod_{n=1}^{\infty}\frac{\left(1-q^{2n}\right)^{2}\left(1-q^{6n}\right)^{14}\left(1-q^{12n}\right)^{20}}{\left(1-q^{4n}\right)^{4}}+r_{14}q^{9}\prod_{n=1}^{\infty}\frac{\left(1-q^{2n}\right)^{16}\left(1-q^{4n}\right)^{10}\left(1-q^{12n}\right)^{18}}{\left(1-q^{6n}\right)^{12}}$$

$$+r_{15}q^{12}\prod_{n=1}^{\infty}\frac{\left(1-q^{4n}\right)^{19}\left(1-q^{6n}\right)^{9}\left(1-q^{12n}\right)^{15}}{\left(1-q^{2n}\right)^{11}}+r_{16}q^{12}\prod_{n=1}^{\infty}\frac{\left(1-q^{4n}\right)^{14}\left(1-q^{6n}\right)^{14}\left(1-q^{12n}\right)^{14}}{\left(1-q^{2n}\right)^{10}}$$

$$+r_{17}q^{14}\prod_{n=1}^{\infty}\frac{\left(1-q^{4n}\right)^{15}\left(1-q^{6n}\right)^{13}\left(1-q^{12n}\right)^{19}}{\left(1-q^{2n}\right)^{15}}+r_{18}q^{12}\prod_{n=1}^{\infty}\frac{\left(1-q^{4n}\right)^{9}\left(1-q^{6n}\right)^{19}\left(1-q^{12n}\right)^{13}}{\left(1-q^{2n}\right)^{9}}$$

$$+r_{19}q^{10}\prod_{n=1}^{\infty}\frac{\left(1-q^{4n}\right)^{18}\left(1-q^{6n}\right)^{10}\left(1-q^{12n}\right)^{10}}{\left(1-q^{2n}\right)^{6}}$$

$$+r_{20}\prod_{n=1}^{\infty}\frac{\left(1-q^{2n}\right)^{20}\left(1-q^{4n}\right)^{8}\left(1-q^{6n}\right)^{20}}{\left(1-q^{12n}\right)^{16}}$$

$$+r_{21}q^{2}\prod_{n=1}^{\infty}\frac{\left(1-q^{2n}\right)^{20}\left(1-q^{4n}\right)^{20}}{\left(1-q^{6n}\right)^{4}\left(1-q^{12n}\right)^{4}}$$

$$+r_{22}q^{12}\prod_{n=1}^{\infty}\frac{\left(1-q^{4n}\right)^{20}\left(1-q^{12n}\right)^{20}}{\left(1-q^{2n}\right)^{4}\left(1-q^{6n}\right)^{4}}$$

$$+r_{23}q^{6}\prod_{n=1}^{\infty}\frac{\left(1-q^{4n}\right)^{20}\left(1-q^{6n}\right)^{20}}{\left(1-q^{2n}\right)^{4}\left(1-q^{12n}\right)^{4}}$$

$$+r_{24}q^{12}\prod_{n=1}^{\infty}\frac{\left(1-q^{4n}\right)^{15}\left(1-q^{6n}\right)\left(1-q^{12n}\right)^{19}}{\left(1-q^{2n}\right)^{3}}$$

$$+r_{25}q^{8}\prod_{n=1}^{\infty}\frac{\left(1-q^{2n}\right)^{20}\left(1-q^{4n}\right)^{20}}{\left(1-q^{6n}\right)^{4}\left(1-q^{12n}\right)^{4}}$$

$$+r_{26}q^{12}\prod_{n=1}^{\infty}\frac{\left(1-q^{2n}\right)^{20}\left(1-q^{6n}\right)^{20}\left(1-q^{12n}\right)^{8}}{\left(1-q^{4n}\right)^{4}}$$

$$+r_{27}\prod_{n=1}^{\infty}\frac{\left(1-q^{2n}\right)^{19}\left(1-q^{4n}\right)^{13}\left(1-q^{6n}\right)^{15}}{\left(1-q^{12n}\right)^{15}}$$

$$=\delta(b_1)-\sum_{n=1}^{\infty}\left(c_1\sigma_{15}(n)+c_2\sigma_{15}\left(\frac{n}{2}\right)+c_3\sigma_{15}\left(\frac{n}{3}\right)+c_4\sigma_{15}\left(\frac{n}{4}\right)\right.$$

$$\left.+c_6\sigma_{15}\left(\frac{n}{6}\right)+c_{12}\sigma_{15}\left(\frac{n}{12}\right)\right)+r_1f_1(n)+\cdots+r_{27}f_{27}(n),$$

where

$$\delta(b_1)=\begin{cases}0 & \text{if } b_1\neq 0\\ 1 & \text{if } b_1=0\end{cases}.$$

So

$$c(n)=-\left(c_1\sigma_{15}(n)+c_2\sigma_{15}\left(\frac{n}{2}\right)+c_3\sigma_{15}\left(\frac{n}{3}\right)+c_4\sigma_{15}\left(\frac{n}{4}\right)\right.$$
$$\left.+c_6\sigma_{15}\left(\frac{n}{6}\right)+c_{12}\sigma_{15}\left(\frac{n}{12}\right)\right)+r_1f_1(n)+\cdots+r_{27}f_{27}(n).$$

Therefore, for $n=1,2,\cdots,$

$$c(2n)=-c_1\sigma_{15}(2n)-c_2\sigma_{15}(n)-c_4\sigma_{15}\left(\frac{n}{2}\right)-(65537c_3+c_6)\sigma_{15}\left(\frac{n}{3}\right)$$
$$-(c_{12}-65536c_3)\sigma_{15}\left(\frac{n}{6}\right)+r_{15}f_{15}(2n)+\cdots+r_{27}f_{27}(2n),$$

$$c(2n-1)=-c_1\sigma_{15}(2n-1)-c_3\sigma_{15}\left(\frac{2n-1}{3}\right)+r_1f_1(2n-1)+\cdots+r_{14}f_{14}(2n-1),$$

since it is easy to see that

$$\sigma_k\left(\frac{2n}{3}\right)=\left(2^k+1\right)\sigma_k\left(\frac{n}{3}\right)-2^k\sigma_k\left(\frac{n}{6}\right)$$

hence,

$$\sigma_{15}\left(\frac{2n}{3}\right)=65537\sigma_{15}\left(\frac{n}{3}\right)-65536\sigma_{15}\left(\frac{n}{6}\right),$$

and, for $n=1,2,\cdots,$

$$f_1(2n)=\cdots=f_{14}(2n)=0,$$

$$f_{15}(2n-1)=\cdots=f_{27}(2n-1)=0.$$

Remark 2 We have found 360 eta quotients, see Table 2, such that, for $n=1,2,\cdots$,

$$c(2n)=-c_1\sigma_{15}(2n)-c_2\sigma_{15}(n)-c_4\sigma_{15}\left(\frac{n}{2}\right)-(65537c_3+c_6)\sigma_{15}\left(\frac{n}{3}\right)$$
$$-(c_{12}-65536c_3)\sigma_{15}\left(\frac{n}{6}\right)$$

$$c(2n-1)=-c_1\sigma_{15}(2n-1)-c_3\sigma_{15}\left(\frac{2n-1}{3}\right)+r_1f_1(2n-1)$$
$$+\cdots+r_{14}f_{14}(2n-1).$$

and 875 eta quotients, such that for $n=1,2,\cdots$,

$$c(2n)=-c_1\sigma_{15}(2n)-c_2\sigma_{15}(n)-c_4\sigma_{15}\left(\frac{n}{2}\right)-c_6\sigma_{15}\left(\frac{n}{3}\right)$$
$$-c_{12}\sigma_{15}\left(\frac{n}{6}\right)+r_{15}f_{15}(2n)+\cdots+r_{27}f_{27}(2n),$$

$$c(2n-1)=0.$$

Remark 3 If f is an eta quotient, then $f(-q)$ is also an eta quotient, so the coefficients of $\frac{1}{2}(f(q)+f(-q))$

are exactly the even coefficients of f. In particular, it means that we have obtained all coefficients of some sum of 360 eta quotients.

Table 2. The eta quotients whose even coefficients can be explicitly calculated.

No	a_3	a_4	a_6	a_{12}	c_1	c_2	c_3	c_4	c_6	c_{12}	No
1	−46	92	14	−28	$\frac{1}{217408}$	$-\frac{32769}{217408}$	0	$\frac{512}{3397}$	0	0	1
2	−45	87	19	−29	$\frac{65}{14348928}$	$-\frac{64545}{434816}$	0	$\frac{16640}{112101}$	0	0	2
3	−44	82	24	−30	$\frac{1}{224202}$	$-\frac{993}{6794}$	0	$\frac{16384}{112101}$	0	0	3
4	−43	77	29	−31	$\frac{21}{4782976}$	$-\frac{62559}{434816}$	0	$\frac{5376}{37367}$	0	0	4
5	−42	72	34	−32	$\frac{31}{7174464}$	$-\frac{30783}{217408}$	0	$\frac{15872}{112101}$	0	0	5
6	−41	67	39	−33	$\frac{61}{14348928}$	$-\frac{60573}{434816}$	0	$\frac{15616}{112101}$	0	0	6
7	−40	62	44	−34	$\frac{5}{1195744}$	$-\frac{14895}{108704}$	0	$\frac{5120}{37367}$	0	0	7
8	−39	57	49	−35	$\frac{59}{14348928}$	$-\frac{58587}{434816}$	0	$\frac{15104}{112101}$	0	0	8
9	−38	52	54	−36	$\frac{29}{7174464}$	$-\frac{28797}{217408}$	0	$\frac{14848}{112101}$	0	0	9
10	−37	47	59	−37	$\frac{19}{4782976}$	$-\frac{56601}{434816}$	0	$\frac{4864}{37367}$	0	0	10
11	−36	42	64	−38	$\frac{7}{1793616}$	$-\frac{6951}{54352}$	0	$\frac{14336}{112101}$	0	0	11
12	−35	37	69	−39	$\frac{5}{1304448}$	$-\frac{54615}{434816}$	0	$\frac{1280}{10191}$	0	0	12
13	−34	32	74	−40	$\frac{9}{2391488}$	$-\frac{26811}{217408}$	0	$\frac{4608}{37367}$	0	0	13
14	−33	27	79	−41	$\frac{53}{14348928}$	$-\frac{52629}{434816}$	0	$\frac{13568}{112101}$	0	0	14
15	−32	22	84	−42	$\frac{13}{3587232}$	$-\frac{12909}{108704}$	0	$\frac{13312}{112101}$	0	0	15
16	−31	17	89	−43	$\frac{6971}{1961319096}$	$-\frac{2307401}{19811304}$	$\frac{1}{2187}$	$\frac{28553216}{245164887}$	$-\frac{3641}{243}$	$\frac{32768}{2187}$	16
17	−37	83	11	−25	$\frac{19}{4782976}$	$-\frac{56601}{434816}$	0	$\frac{4864}{37367}$	0	0	17
18	−36	78	16	−26	$\frac{7}{1793616}$	$-\frac{6951}{54352}$	0	$\frac{14336}{112101}$	0	0	18
19	−35	73	21	−27	$\frac{5}{1304448}$	$-\frac{54615}{434816}$	0	$\frac{1280}{10191}$	0	0	19
20	−34	68	26	−28	$\frac{9}{2391488}$	$-\frac{26811}{217408}$	0	$\frac{4608}{37367}$	0	0	20
21	−33	63	31	−29	$\frac{53}{14348928}$	$-\frac{52629}{434816}$	0	$\frac{13568}{112101}$	0	0	21

22	−32	58	36	−30	$\frac{13}{3587232}$	$-\frac{12909}{108704}$	0	$\frac{13312}{112101}$	0	0	22
23	−31	53	41	−31	$\frac{17}{4782976}$	$-\frac{50643}{434816}$	0	$\frac{4352}{37367}$	0	0	23
24	−30	48	46	−32	$\frac{25}{7174464}$	$-\frac{24825}{217408}$	0	$\frac{12800}{112101}$	0	0	24
25	−29	43	51	−33	$\frac{49}{14348928}$	$-\frac{48657}{434816}$	0	$\frac{12544}{112101}$	0	0	25
26	−28	38	56	−34	$\frac{1}{298936}$	$-\frac{2979}{27176}$	0	$\frac{4096}{37367}$	0	0	26
27	−27	33	61	−35	$\frac{47}{14348928}$	$-\frac{46671}{434816}$	0	$\frac{12032}{112101}$	0	0	27
28	−26	28	66	−36	$\frac{23}{7174464}$	$-\frac{22839}{217408}$	0	$\frac{11776}{112101}$	0	0	28
29	−25	23	71	−37	$\frac{15}{4782976}$	$-\frac{44685}{434816}$	0	$\frac{3840}{37367}$	0	0	29
30	−24	18	76	−38	$\frac{1}{326112}$	$-\frac{10923}{108704}$	0	$\frac{1024}{10191}$	0	0	30
31	−23	13	81	−39	$\frac{653}{217924344}$	$-\frac{216143}{2201256}$	$\frac{1}{243}$	$\frac{2674688}{27240543}$	$-\frac{3641}{27}$	$\frac{32768}{243}$	31
32	−28	74	8	−22	$\frac{1}{298936}$	$-\frac{2979}{27176}$	0	$\frac{4096}{37367}$	0	0	32
33	−27	69	13	−23	$\frac{47}{14348928}$	$-\frac{46671}{434816}$	0	$\frac{12032}{112101}$	0	0	33
34	−26	64	18	−24	$\frac{23}{7174464}$	$-\frac{22839}{217408}$	0	$\frac{11776}{112101}$	0	0	34
35	−25	59	23	−25	$\frac{15}{4782976}$	$-\frac{44685}{434816}$	0	$\frac{3840}{37367}$	0	0	35
36	−24	54	28	−26	$\frac{1}{326112}$	$-\frac{10923}{108704}$	0	$\frac{1024}{10191}$	0	0	36
37	−23	49	33	−27	$\frac{1}{333696}$	$-\frac{993}{10112}$	0	$\frac{256}{2607}$	0	0	37
38	−22	44	38	−28	$\frac{7}{2391488}$	$-\frac{20853}{217408}$	0	$\frac{3584}{37367}$	0	0	38
39	−21	39	43	−29	$\frac{41}{14348928}$	$-\frac{40713}{434816}$	0	$\frac{10496}{112101}$	0	0	39
40	−20	34	48	−30	$\frac{5}{1793616}$	$-\frac{4965}{54352}$	0	$\frac{10240}{112101}$	0	0	40
41	−19	29	53	−31	$\frac{13}{4782976}$	$-\frac{38727}{434816}$	0	$\frac{3328}{37367}$	0	0	41
42	−18	24	58	−32	$\frac{19}{7174464}$	$-\frac{18867}{217408}$	0	$\frac{9728}{112101}$	0	0	42
43	−17	19	63	−33	$\frac{37}{14348928}$	$-\frac{36741}{434816}$	0	$\frac{9472}{112101}$	0	0	43

44	−16	14	68	−34	$\frac{3}{1195744}$	$-\frac{8937}{108704}$	0	$\frac{3072}{37367}$	0	0	44
45	−15	9	73	−35	$\frac{59}{24213816}$	$-\frac{19529}{244584}$	$\frac{1}{27}$	$\frac{241664}{3026727}$	$-\frac{3641}{3}$	$\frac{32768}{27}$	45
46	−19	65	5	−19	$\frac{13}{4782976}$	$-\frac{38727}{434816}$	0	$\frac{3328}{37367}$	0	0	46
47	−18	60	10	−20	$\frac{19}{7174464}$	$-\frac{18867}{217408}$	0	$\frac{9728}{112101}$	0	0	47
48	−17	55	15	−21	$\frac{37}{14348928}$	$-\frac{36741}{434816}$	0	$\frac{9472}{112101}$	0	0	48
49	−16	50	20	−22	$\frac{3}{1195744}$	$-\frac{8937}{108704}$	0	$\frac{3072}{37367}$	0	0	49
50	−15	45	25	−23	$\frac{35}{14348928}$	$-\frac{34755}{434816}$	0	$\frac{8960}{112101}$	0	0	50
51	−14	40	30	−24	$\frac{17}{7174464}$	$-\frac{16881}{217408}$	0	$\frac{8704}{112101}$	0	0	51
52	−13	35	35	−25	$\frac{1}{434816}$	$-\frac{32769}{434816}$	0	$\frac{256}{3397}$	0	0	52
53	−12	30	40	−26	$\frac{1}{448404}$	$-\frac{993}{13588}$	0	$\frac{8192}{112101}$	0	0	53
54	−11	25	45	−27	$\frac{31}{14348928}$	$-\frac{30783}{434816}$	0	$\frac{7936}{112101}$	0	0	54
55	−10	20	50	−28	$\frac{5}{2391488}$	$-\frac{14895}{217408}$	0	$\frac{2560}{37367}$	0	0	55
56	−9	15	55	−29	$\frac{29}{14348928}$	$-\frac{28797}{434816}$	0	$\frac{7424}{112101}$	0	0	56
57	−8	10	60	−30	$\frac{7}{3587232}$	$-\frac{6951}{108704}$	0	$\frac{7168}{112101}$	0	0	57
58	−7	5	65	−31	$\frac{5}{2690424}$	$-\frac{1655}{27176}$	$\frac{1}{3}$	$\frac{20480}{336303}$	−10923	$\frac{32768}{3}$	58
59	−10	56	2	−16	$\frac{5}{2391488}$	$-\frac{14895}{217408}$	0	$\frac{2560}{37367}$	0	0	59
60	−9	51	7	−17	$\frac{29}{14348928}$	$-\frac{28797}{434816}$	0	$\frac{7424}{112101}$	0	0	60
61	−8	46	12	−18	$\frac{7}{3587232}$	$-\frac{6951}{108704}$	0	$\frac{7168}{112101}$	0	0	61
62	−7	41	17	−19	$\frac{9}{4782976}$	$-\frac{26811}{434816}$	0	$\frac{2304}{37367}$	0	0	62
63	−6	36	22	−20	$\frac{13}{7174464}$	$-\frac{12909}{217408}$	0	$\frac{6656}{112101}$	0	0	63
64	−5	31	27	−21	$\frac{25}{14348918}$	$-\frac{24825}{434816}$	0	$\frac{6400}{112101}$	0	0	64
65	−4	26	32	−22	$\frac{1}{597872}$	$-\frac{2979}{54352}$	0	$\frac{2048}{37367}$	0	0	65

66	−3	21	37	−23	$\frac{23}{14348928}$	$-\frac{22839}{434816}$	0	$\frac{5888}{112101}$	0	0	66
67	−2	16	42	−24	$\frac{1}{652224}$	$-\frac{10923}{217408}$	0	$\frac{512}{10191}$	0	0	67
68	−1	11	47	−25	$\frac{7}{4782976}$	$-\frac{20853}{434816}$	0	$\frac{1792}{37367}$	0	0	68
69	0	6	52	−26	$\frac{5}{3587232}$	$-\frac{4965}{108704}$	0	$\frac{5120}{112101}$	0	0	69
70	1	1	57	−27	$\frac{1}{896808}$	$-\frac{993}{27176}$	3	$\frac{4096}{112101}$	−98307	98304	70
71	−1	47	−1	−13	$\frac{7}{4782976}$	$-\frac{20853}{434816}$	0	$\frac{1792}{37367}$	0	0	71
72	0	42	4	−14	$\frac{5}{3587232}$	$-\frac{4965}{108704}$	0	$\frac{5120}{112101}$	0	0	72
73	1	37	9	−15	$\frac{19}{14348928}$	$-\frac{18867}{434816}$	0	$\frac{4864}{112101}$	0	0	73
74	2	32	14	−16	$\frac{3}{2391488}$	$-\frac{8937}{217408}$	0	$\frac{1536}{37367}$	0	0	74
75	3	27	19	−17	$\frac{17}{14348928}$	$-\frac{16881}{434816}$	0	$\frac{4352}{112101}$	0	0	75
76	4	22	24	−18	$\frac{1}{896808}$	$-\frac{993}{27176}$	0	$\frac{4096}{112101}$	0	0	76
77	5	17	29	−19	$\frac{5}{4782976}$	$-\frac{14895}{434816}$	0	$\frac{1280}{37367}$	0	0	77
78	6	12	34	−20	$\frac{7}{7174464}$	$-\frac{6951}{217408}$	0	$\frac{3584}{112101}$	0	0	78
79	7	7	39	−21	$\frac{13}{14348928}$	$-\frac{12909}{434816}$	0	$\frac{3328}{112101}$	0	0	79
80	8	2	44	−22	$\frac{1}{1195744}$	$-\frac{2979}{108704}$	0	$\frac{1024}{37367}$	0	0	80
81	9	−3	49	−23	$-\frac{1}{896808}$	$\frac{993}{27176}$	27	$-\frac{4096}{112101}$	−884763	884736	81
82	8	38	−4	−10	$\frac{1}{1195744}$	$-\frac{2979}{108704}$	0	$\frac{1024}{37367}$	0	0	82
83	9	33	1	−11	$\frac{1}{1304448}$	$-\frac{10923}{434816}$	0	$\frac{256}{10191}$	0	0	83
84	10	28	6	−12	$\frac{5}{7174464}$	$-\frac{4965}{217408}$	0	$\frac{2560}{112101}$	0	0	84
85	11	23	11	−13	$\frac{3}{4782976}$	$-\frac{8937}{434816}$	0	$\frac{768}{37367}$	0	0	85
86	12	18	16	−14	$\frac{1}{1793616}$	$-\frac{993}{54352}$	0	$\frac{2048}{112101}$	0	0	86
87	13	13	21	−15	$\frac{7}{14348928}$	$-\frac{6951}{434816}$	0	$\frac{1792}{112101}$	0	0	87

88	14	8	26	−16	$\frac{1}{2391488}$	$-\frac{2979}{217408}$	0	$\frac{512}{37367}$	0	0	88
89	15	3	31	−17	$\frac{5}{14348928}$	$-\frac{4965}{434816}$	0	$\frac{1280}{112101}$	0	0	89
90	16	−2	36	−18	$\frac{1}{3587232}$	$-\frac{993}{108704}$	0	$\frac{1024}{112101}$	0	0	90
91	17	−7	41	−19	$-\frac{5}{298936}$	$\frac{14895}{27176}$	243	$-\frac{20480}{37367}$	−7962867	7962624	91
92	17	29	−7	−7	$\frac{1}{4782976}$	$-\frac{2979}{434816}$	0	$\frac{256}{37367}$	0	0	92
93	18	24	−2	−8	$\frac{1}{7174464}$	$-\frac{993}{217408}$	0	$\frac{512}{112101}$	0	0	93
94	19	19	3	−9	$\frac{1}{14348928}$	$-\frac{993}{434816}$	0	$\frac{256}{112101}$	0	0	94
95	21	9	13	−11	$-\frac{1}{14348928}$	$\frac{993}{434816}$	0	$-\frac{256}{112101}$	0	0	95
96	22	4	18	−12	$-\frac{1}{7174464}$	$\frac{993}{217408}$	0	$-\frac{512}{112101}$	0	0	96
97	23	−1	23	−13	$-\frac{1}{4782976}$	$\frac{2979}{434816}$	0	$-\frac{256}{37367}$	0	0	97
98	24	−6	28	−14	$-\frac{1}{3587232}$	$\frac{993}{108704}$	0	$-\frac{1024}{112101}$	0	0	98
99	25	−11	33	−15	$-\frac{137}{896808}$	$\frac{136041}{27176}$	2187	$-\frac{561152}{112101}$	−71665803	71663616	99
100	26	20	−10	−4	$-\frac{1}{2391488}$	$\frac{2979}{217408}$	0	$-\frac{512}{37367}$	0	0	100
101	27	15	−5	−5	$-\frac{7}{14348928}$	$\frac{6951}{434816}$	0	$-\frac{1792}{112101}$	0	0	101
102	28	10	0	−6	$-\frac{1}{1793616}$	$\frac{993}{54352}$	0	$-\frac{2048}{112101}$	0	0	102
103	29	5	5	−7	$-\frac{3}{4782976}$	$\frac{8937}{434816}$	0	$-\frac{768}{37367}$	0	0	103
104	30	0	10	−8	$-\frac{5}{7174464}$	$\frac{4965}{217408}$	0	$-\frac{2560}{112101}$	0	0	104
105	31	−5	15	−9	$-\frac{1}{1304448}$	$\frac{10923}{434816}$	0	$-\frac{256}{10191}$	0	0	105
106	32	−10	20	−10	$-\frac{1}{1195744}$	$\frac{2979}{108704}$	0	$-\frac{1024}{37367}$	0	0	106
107	33	−15	25	−11	$-\frac{1231}{896808}$	$\frac{1222383}{27176}$	19683	$-\frac{5042176}{112101}$	−644992227	644972544	107
108	35	11	−13	−1	$-\frac{5}{4782976}$	$\frac{14895}{434816}$	0	$-\frac{1280}{37367}$	0	0	108
109	36	6	−8	2	$-\frac{1}{896808}$	$\frac{993}{27176}$	0	$-\frac{4096}{112101}$	0	0	109

110	37	1	−3	−3	$-\frac{17}{14348928}$	$\frac{16881}{434816}$	0	$-\frac{4352}{112101}$	0	0	110
111	38	−4	2	−4	$-\frac{3}{2391488}$	$\frac{8937}{217408}$	0	$-\frac{1536}{37367}$	0	0	111
112	39	−9	7	−5	$-\frac{19}{14348928}$	$\frac{18867}{434816}$	0	$-\frac{4864}{112101}$	0	0	112
113	40	−14	12	−6	$-\frac{5}{3587232}$	$\frac{4965}{108704}$	0	$-\frac{5120}{112101}$	0	0	113
114	41	−19	17	−7	$-\frac{3691}{298936}$	$\frac{10995489}{27176}$	177147	$-\frac{15118336}{37367}$	−5804930043	5804752896	114
115	44	2	−16	2	$-\frac{1}{597872}$	$\frac{2979}{54352}$	0	$-\frac{2048}{37367}$	0	0	115
116	45	−3	−11	1	$-\frac{25}{14348918}$	$\frac{24825}{434816}$	0	$-\frac{6400}{112101}$	0	0	116
117	46	−8	−6	0	$-\frac{13}{7174464}$	$\frac{12909}{217408}$	0	$-\frac{6656}{112101}$	0	0	117
118	47	−13	−1	−1	$-\frac{9}{4782976}$	$\frac{26811}{434816}$	0	$-\frac{2304}{37367}$	0	0	118
119	48	−18	4	−2	$-\frac{7}{3587232}$	$\frac{6951}{108704}$	0	$-\frac{7168}{112101}$	0	0	119
120	49	−23	9	−3	$-\frac{99647}{896808}$	$\frac{98949471}{27176}$	1594323	$-\frac{408154112}{112101}$	−52244370387	52242776064	120
121	53	−7	−19	5	$-\frac{1}{434816}$	$\frac{32769}{434816}$	0	$-\frac{256}{3397}$	0	0	121
122	54	−12	−14	4	$-\frac{17}{7174464}$	$\frac{16881}{217408}$	0	$-\frac{8704}{112101}$	0	0	122
123	55	−17	−9	3	$-\frac{35}{14348928}$	$\frac{34755}{434816}$	0	$-\frac{8960}{112101}$	0	0	123
124	56	−22	−4	2	$-\frac{3}{1195744}$	$\frac{8937}{108704}$	0	$-\frac{3072}{37367}$	0	0	124
125	57	−27	1	1	$-\frac{896809}{896808}$	$\frac{890531337}{27176}$	14348907	$-\frac{3673329664}{112101}$	−470199333483	470184984576	125
126	62	−16	−22	8	$-\frac{7}{2391488}$	$\frac{20853}{217408}$	0	$-\frac{3584}{37367}$	0	0	126
127	63	−21	−17	7	$-\frac{1}{333696}$	$\frac{993}{10112}$	0	$-\frac{256}{2607}$	0	0	127
128	64	−26	−12	6	$-\frac{1}{326112}$	$\frac{10923}{108704}$	0	$-\frac{1024}{10191}$	0	0	128
129	65	−31	−7	5	$-\frac{2690421}{298936}$	$\frac{8014764159}{27176}$	129140163	$-\frac{11019964416}{37367}$	−4231794001347	4231664861184	129
130	71	−25	−25	11	$-\frac{17}{4782976}$	$\frac{50643}{434816}$	0	$-\frac{4352}{37367}$	0	0	130
131	72	−30	−20	10	$\frac{13}{3587232}$	$-\frac{12909}{108704}$	0	$\frac{13312}{112101}$	0	0	131

132	73	−35	−15	9	$-\frac{72641345}{896808}$	$\frac{72132855585}{27176}$	1162261467	$-\frac{297538949120}{112101}$	−38086146012123	38084983750656	132
133	80	−34	−28	14	$-\frac{5}{1195744}$	$\frac{14895}{108704}$	0	$-\frac{5120}{37367}$	0	0	133
134	81	−39	−23	13	$-\frac{653772079}{896808}$	$\frac{649195674447}{27176}$	10460353203	$-\frac{2677850435584}{112101}$	−342775314109107	342764853755904	134
135	89	−43	−31	17	$-\frac{1961316227}{298936}$	$\frac{5842761040233}{27176}$	94143178827	$-\frac{8033551265792}{37367}$	−3084977826981963	3084883683803136	135
136	−42	84	10	−20	$\frac{1}{14348928}$	$-\frac{993}{434816}$	0	$\frac{256}{112101}$	0	0	136
137	−41	79	15	−21	$\frac{1}{14348928}$	$-\frac{993}{434816}$	0	$\frac{256}{112101}$	0	0	137
138	−40	74	20	−22	$\frac{1}{14348928}$	$-\frac{993}{434816}$	0	$\frac{256}{112101}$	0	0	138
139	−39	69	25	−23	$\frac{1}{14348928}$	$-\frac{993}{434816}$	0	$\frac{256}{112101}$	0	0	139
140	−38	64	30	−24	$\frac{1}{14348928}$	$-\frac{993}{434816}$	0	$\frac{256}{112101}$	0	0	140
141	−37	59	35	−25	$\frac{1}{14348928}$	$-\frac{993}{434816}$	0	$\frac{256}{112101}$	0	0	141
142	−36	54	40	−26	$\frac{1}{14348928}$	$-\frac{993}{434816}$	0	$\frac{256}{112101}$	0	0	142
143	−35	49	45	−27	$\frac{1}{14348928}$	$-\frac{993}{434816}$	0	$\frac{256}{112101}$	0	0	143
144	−34	44	50	−28	$\frac{1}{14348928}$	$-\frac{993}{434816}$	0	$\frac{256}{112101}$	0	0	144
145	−33	39	55	−29	$\frac{1}{14348928}$	$-\frac{993}{434816}$	0	$\frac{256}{112101}$	0	0	145
146	−32	34	60	−30	$\frac{1}{14348928}$	$-\frac{993}{434816}$	0	$\frac{256}{112101}$	0	0	146
147	−31	29	65	−31	$\frac{1}{14348928}$	$-\frac{993}{434816}$	0	$\frac{256}{112101}$	0	0	147
148	−30	24	70	−32	$\frac{1}{14348928}$	$-\frac{993}{434816}$	0	$\frac{256}{112101}$	0	0	148
149	−29	19	75	−33	$\frac{1}{14348928}$	$-\frac{993}{434816}$	0	$\frac{256}{112101}$	0	0	149
150	−28	14	80	−34	$\frac{547}{7845276384}$	$-\frac{181057}{79245216}$	$-\frac{1}{2187}$	$\frac{560128}{245164887}$	$\frac{3641}{243}$	$-\frac{32768}{2187}$	150
151	−33	75	7	−17	$\frac{1}{14348928}$	$-\frac{993}{434816}$	0	$\frac{256}{112101}$	0	0	151
152	−32	70	12	−18	$\frac{1}{14348928}$	$-\frac{993}{434816}$	0	$\frac{256}{112101}$	0	0	152
153	−31	65	17	−19	$\frac{1}{14348928}$	$-\frac{993}{434816}$	0	$\frac{256}{112101}$	0	0	153

154	−30	60	22	−20	$\frac{1}{14348928}$	$-\frac{993}{434816}$	0	$\frac{256}{112101}$	0	0	154
155	−29	55	27	−21	$\frac{1}{14348928}$	$-\frac{993}{434816}$	0	$\frac{256}{112101}$	0	0	155
156	−28	50	32	−22	$\frac{1}{14348928}$	$-\frac{993}{434816}$	0	$\frac{256}{112101}$	0	0	156
157	−27	45	37	−23	$\frac{1}{14348928}$	$-\frac{993}{434816}$	0	$\frac{256}{112101}$	0	0	157
158	−26	40	42	−24	$\frac{1}{14348928}$	$-\frac{993}{434816}$	0	$\frac{256}{112101}$	0	0	158
159	−25	35	47	−25	$\frac{1}{14348928}$	$-\frac{993}{434816}$	0	$\frac{256}{112101}$	0	0	159
160	−24	30	52	−26	$\frac{1}{14348928}$	$-\frac{993}{434816}$	0	$\frac{256}{112101}$	0	0	160
161	−23	25	57	−27	$\frac{1}{14348928}$	$-\frac{993}{434816}$	0	$\frac{256}{112101}$	0	0	161
162	−22	20	62	−28	$\frac{1}{14348928}$	$-\frac{993}{434816}$	0	$\frac{256}{112101}$	0	0	162
163	−21	15	67	−29	$\frac{1}{14348928}$	$-\frac{993}{434816}$	0	$\frac{256}{112101}$	0	0	163
164	−20	10	72	−30	$\frac{61}{871697376}$	$-\frac{20191}{8805024}$	$-\frac{1}{243}$	$\frac{62464}{27240543}$	$\frac{3641}{27}$	$-\frac{32768}{243}$	164
165	−24	66	4	−14	$\frac{1}{14348928}$	$-\frac{993}{434816}$	0	$\frac{256}{112101}$	0	0	165
166	−23	61	9	−15	$\frac{1}{14348928}$	$-\frac{993}{434816}$	0	$\frac{256}{112101}$	0	0	166
167	−22	56	14	−16	$\frac{1}{14348928}$	$-\frac{993}{434816}$	0	$\frac{256}{112101}$	0	0	167
168	−21	51	19	−17	$\frac{1}{14348928}$	$-\frac{993}{434816}$	0	$\frac{256}{112101}$	0	0	168
169	−20	46	24	−18	$\frac{1}{14348928}$	$-\frac{993}{434816}$	0	$\frac{256}{112101}$	0	0	169
170	−19	41	29	−19	$\frac{1}{14348928}$	$-\frac{993}{434816}$	0	$\frac{256}{112101}$	0	0	170
171	−18	36	34	−20	$\frac{1}{14348928}$	$-\frac{993}{434816}$	0	$\frac{256}{112101}$	0	0	171
172	−17	31	39	−21	$\frac{1}{14348928}$	$-\frac{993}{434816}$	0	$\frac{256}{112101}$	0	0	172
173	−16	26	44	−22	$\frac{1}{14348928}$	$-\frac{993}{434816}$	0	$\frac{256}{112101}$	0	0	173
174	−15	21	49	−23	$\frac{1}{14348928}$	$-\frac{993}{434816}$	0	$\frac{256}{112101}$	0	0	174
175	−14	16	54	−24	$\frac{1}{14348928}$	$-\frac{993}{434816}$	0	$\frac{256}{112101}$	0	0	175

176	−13	11	59	−25	$\frac{1}{14348928}$	$-\frac{993}{434816}$	0	$\frac{256}{112101}$	0	0	176
177	−12	6	64	−26	$\frac{7}{96855264}$	$-\frac{2317}{978336}$	$-\frac{1}{27}$	$\frac{7168}{3026727}$	$\frac{3641}{3}$	$-\frac{32768}{27}$	177
178	−15	57	1	−11	$\frac{1}{14348928}$	$-\frac{993}{434816}$	0	$\frac{256}{112101}$	0	0	178
179	−14	52	6	−12	$\frac{1}{14348928}$	$-\frac{993}{434816}$	0	$\frac{256}{112101}$	0	0	179
180	−13	47	11	−13	$\frac{1}{14348928}$	$-\frac{993}{434816}$	0	$\frac{256}{112101}$	0	0	180
181	−12	42	16	−14	$\frac{1}{14348928}$	$-\frac{993}{434816}$	0	$\frac{256}{112101}$	0	0	181
182	−11	37	21	−15	$\frac{1}{14348928}$	$-\frac{993}{434816}$	0	$\frac{256}{112101}$	0	0	182
183	−10	32	26	−16	$\frac{1}{14348928}$	$-\frac{993}{434816}$	0	$\frac{256}{112101}$	0	0	183
184	−9	27	31	−17	$\frac{1}{14348928}$	$-\frac{993}{434816}$	0	$\frac{256}{112101}$	0	0	184
185	−8	22	36	−18	$\frac{1}{14348928}$	$-\frac{993}{434816}$	0	$\frac{256}{112101}$	0	0	185
186	−7	17	41	−19	$\frac{1}{14348928}$	$-\frac{993}{434816}$	0	$\frac{256}{112101}$	0	0	186
187	−6	12	46	−20	$\frac{1}{14348928}$	$-\frac{993}{434816}$	0	$\frac{256}{112101}$	0	0	187
188	−5	7	51	−21	$\frac{1}{14348928}$	$-\frac{993}{434816}$	0	$\frac{256}{112101}$	0	0	188
189	−4	2	56	−22	$\frac{1}{10761696}$	$-\frac{331}{108704}$	$-\frac{1}{3}$	$\frac{1024}{336303}$	10923	$-\frac{32768}{3}$	189
190	−6	48	−2	−8	$\frac{1}{14348928}$	$-\frac{993}{434816}$	0	$\frac{256}{112101}$	0	0	190
191	−5	43	3	−9	$\frac{1}{14348928}$	$-\frac{993}{434816}$	0	$\frac{256}{112101}$	0	0	191
192	−4	38	8	−9	$\frac{1}{14348928}$	$-\frac{993}{434816}$	0	$\frac{256}{112101}$	0	0	192
193	−3	33	13	−11	$\frac{1}{14348928}$	$-\frac{993}{434816}$	0	$\frac{256}{112101}$	0	0	193
194	−2	28	18	−12	$\frac{1}{14348928}$	$-\frac{993}{434816}$	0	$\frac{256}{112101}$	0	0	194
195	−1	23	23	−13	$\frac{1}{14348928}$	$-\frac{993}{434816}$	0	$\frac{256}{112101}$	0	0	195
196	0	18	28	−14	$\frac{1}{14348928}$	$-\frac{993}{434816}$	0	$\frac{256}{112101}$	0	0	196
197	1	13	33	−15	$\frac{1}{14348928}$	$-\frac{993}{434816}$	0	$\frac{256}{112101}$	0	0	197

198	2	8	38	−16	$\frac{1}{14348928}$	$-\frac{993}{434816}$	0	$\frac{256}{112101}$	0	0	198
199	3	3	43	−17	$\frac{1}{14348928}$	$-\frac{993}{434816}$	0	$\frac{256}{112101}$	0	0	199
200	4	−2	48	−18	$\frac{1}{3587232}$	$-\frac{993}{108704}$	−3	$\frac{1024}{112101}$	98307	−98304	200
201	3	39	−5	−5	$\frac{1}{14348928}$	$-\frac{993}{434816}$	0	$\frac{256}{112101}$	0	0	201
202	4	34	0	−6	$\frac{1}{14348928}$	$-\frac{993}{434816}$	0	$\frac{256}{112101}$	0	0	202
203	5	29	5	−7	$\frac{1}{14348928}$	$-\frac{993}{434816}$	0	$\frac{256}{112101}$	0	0	203
204	6	24	10	−8	$\frac{1}{14348928}$	$-\frac{993}{434816}$	0	$\frac{256}{112101}$	0	0	204
205	7	19	15	−9	$\frac{1}{14348928}$	$-\frac{993}{434816}$	0	$\frac{256}{112101}$	0	0	205
206	8	14	20	−10	$\frac{1}{14348928}$	$-\frac{993}{434816}$	0	$\frac{256}{112101}$	0	0	206
207	9	9	25	−11	$\frac{1}{14348928}$	$-\frac{993}{434816}$	0	$\frac{256}{112101}$	0	0	207
208	10	4	30	−12	$\frac{1}{14348928}$	$-\frac{993}{434816}$	0	$\frac{256}{112101}$	0	0	208
209	11	−1	35	−13	$\frac{1}{14348928}$	$-\frac{993}{434816}$	0	$\frac{256}{112101}$	0	0	209
210	12	−6	40	−14	$\frac{7}{3587232}$	$-\frac{6951}{108704}$	−27	$\frac{7168}{112101}$	884763	−884736	210
211	12	30	−8	−2	$\frac{1}{14348928}$	$-\frac{993}{434816}$	0	$\frac{256}{112101}$	0	0	211
212	13	25	−3	−3	$\frac{1}{14348928}$	$-\frac{993}{434816}$	0	$\frac{256}{112101}$	0	0	212
213	14	20	2	−4	$\frac{1}{14348928}$	$-\frac{993}{434816}$	0	$\frac{256}{112101}$	0	0	213
214	15	15	7	−5	$\frac{1}{14348928}$	$-\frac{993}{434816}$	0	$\frac{256}{112101}$	0	0	214
215	16	10	12	−6	$\frac{1}{14348928}$	$-\frac{993}{434816}$	0	$\frac{256}{112101}$	0	0	215
216	17	5	17	−7	$\frac{1}{14348928}$	$-\frac{993}{434816}$	0	$\frac{256}{112101}$	0	0	216
217	18	0	22	−8	$\frac{1}{14348928}$	$-\frac{993}{434816}$	0	$\frac{256}{112101}$	0	0	217
218	19	−5	27	−9	$\frac{1}{14348928}$	$-\frac{993}{434816}$	0	$\frac{256}{112101}$	0	0	218
219	20	−10	32	−10	$\frac{61}{3587232}$	$-\frac{60573}{108704}$	−243	$\frac{62464}{112101}$	7962867	−7962624	219

220	21	21	−11	1	$\frac{1}{14348928}$	$-\frac{993}{434816}$	0	$\frac{256}{112101}$	0	0	220
221	22	16	−6	0	$\frac{1}{14348928}$	$-\frac{993}{434816}$	0	$\frac{256}{112101}$	0	0	221
222	23	11	−1	−1	$\frac{1}{14348928}$	$-\frac{993}{434816}$	0	$\frac{256}{112101}$	0	0	222
223	24	6	4	−2	$\frac{1}{14348928}$	$-\frac{993}{434816}$	0	$\frac{256}{112101}$	0	0	223
224	25	1	9	−3	$\frac{1}{14348928}$	$-\frac{993}{434816}$	0	$\frac{256}{112101}$	0	0	224
225	26	−4	14	−4	$\frac{1}{14348928}$	$-\frac{993}{434816}$	0	$\frac{256}{112101}$	0	0	225
226	27	−9	19	−5	$\frac{1}{14348928}$	$-\frac{993}{434816}$	0	$\frac{256}{112101}$	0	0	226
227	28	−14	24	−6	$\frac{547}{3587232}$	$-\frac{543171}{108704}$	−2187	$\frac{560128}{112101}$	71665803	−71663616	227
228	30	12	−14	4	$\frac{1}{14348928}$	$-\frac{993}{434816}$	0	$\frac{256}{112101}$	0	0	228
229	31	7	−9	3	$\frac{1}{14348928}$	$-\frac{993}{434816}$	0	$\frac{256}{112101}$	0	0	229
230	32	2	−4	2	$\frac{1}{14348928}$	$-\frac{993}{434816}$	0	$\frac{256}{112101}$	0	0	230
231	33	−3	1	1	$\frac{1}{14348928}$	$-\frac{993}{434816}$	0	$\frac{256}{112101}$	0	0	231
232	34	−8	6	0	$\frac{1}{14348928}$	$-\frac{993}{434816}$	0	$\frac{256}{112101}$	0	0	232
233	35	−13	11	−1	$\frac{1}{14348928}$	$-\frac{993}{434816}$	0	$\frac{256}{112101}$	0	0	233
234	36	−18	16	−2	$\frac{4921}{3587232}$	$-\frac{4886553}{108704}$	−19683	$\frac{5039104}{112101}$	644992227	−644972544	234
235	39	3	−17	7	$\frac{1}{14348928}$	$-\frac{993}{434816}$	0	$\frac{256}{112101}$	0	0	235
236	40	−2	−12	6	$\frac{1}{14348928}$	$-\frac{993}{434816}$	0	$\frac{256}{112101}$	0	0	236
237	41	−7	−7	5	$\frac{1}{14348928}$	$-\frac{993}{434816}$	0	$\frac{256}{112101}$	0	0	237
238	42	−12	−2	4	$\frac{1}{14348928}$	$-\frac{993}{434816}$	0	$\frac{256}{112101}$	0	0	238
239	43	−17	3	3	$\frac{1}{14348928}$	$-\frac{993}{434816}$	0	$\frac{256}{112101}$	0	0	239
240	44	−22	8	2	$\frac{44287}{3587232}$	$-\frac{43976991}{108704}$	−177147	$\frac{45349888}{112101}$	5804930043	−5804752896	240
241	48	−16	−20	10	$\frac{1}{14348928}$	$-\frac{993}{434816}$	0	$\frac{256}{112101}$	0	0	241

242	49	−11	−15	9	$\frac{1}{14348928}$	$-\frac{993}{434816}$	0	$\frac{256}{112101}$	0	0	242
243	50	−16	−10	8	$\frac{1}{14348928}$	$-\frac{993}{434816}$	0	$\frac{256}{112101}$	0	0	243
244	51	−21	−5	7	$\frac{1}{14348928}$	$-\frac{993}{434816}$	0	$\frac{256}{112101}$	0	0	244
245	52	−26	0	6	$\frac{398581}{3587232}$	$-\frac{395790933}{108704}$	−1594323	$\frac{408146944}{112101}$	52244370387	−52242776064	245
246	57	−15	−23	13	$\frac{1}{14348928}$	$-\frac{993}{434816}$	0	$\frac{256}{112101}$	0	0	246
247	58	−20	−18	12	$\frac{1}{14348928}$	$-\frac{993}{434816}$	0	$\frac{256}{112101}$	0	0	247
248	59	−25	−13	11	$\frac{1}{14348928}$	$-\frac{993}{434816}$	0	$\frac{256}{112101}$	0	0	248
249	60	−30	−8	10	$\frac{3587227}{3587232}$	$-\frac{3562116411}{108704}$	−14348907	$\frac{3673320448}{112101}$	470199333483	−470184984576	249
250	66	−24	−26	16	$\frac{1}{14348928}$	$-\frac{993}{434816}$	0	$\frac{256}{112101}$	0	0	250
251	68	−29	−21	15	$\frac{1}{14348928}$	$-\frac{993}{434816}$	0	$\frac{256}{112101}$	0	0	251
252	68	−34	−16	14	$\frac{32285041}{3587232}$	$-\frac{32059045713}{108704}$	−129140163	$\frac{33059881984}{112101}$	4231794001347	−4231664861184	252
253	75	−33	−29	19	$\frac{1}{14348928}$	$-\frac{993}{434816}$	0	$\frac{256}{112101}$	0	0	253
254	76	−38	−24	18	$\frac{290565367}{3587232}$	$-\frac{288531409431}{108704}$	−1162261467	$\frac{297538935808}{112101}$	38086146012123	−38084983750656	254
255	84	−42	−32	22	$\frac{60816007}{83424}$	$-\frac{60390294951}{2528}$	−10460353203	$\frac{62275591168}{2607}$	342775314109107	−342764853755904	255
256	−25	11	71	−25	$-\frac{1}{31381105536}$	$\frac{331}{316980864}$	$\frac{1}{2187}$	$-\frac{256}{245164887}$	$-\frac{3641}{243}$	$\frac{32768}{2187}$	256
257	−17	7	63	−21	$-\frac{1}{3486789504}$	$\frac{331}{35220096}$	$\frac{1}{243}$	$-\frac{256}{27240543}$	$-\frac{3641}{27}$	$\frac{32768}{243}$	257
258	−9	3	55	−17	$-\frac{1}{387421056}$	$\frac{331}{3913344}$	$\frac{1}{27}$	$-\frac{256}{3026727}$	$-\frac{3641}{3}$	$\frac{32768}{27}$	258
259	−1	−1	47	−13	$-\frac{1}{43046784}$	$\frac{331}{434816}$	$\frac{1}{3}$	$-\frac{256}{336303}$	−10923	$\frac{32768}{3}$	259
260	7	−5	39	−9	$-\frac{1}{4782976}$	$\frac{2979}{434816}$	3	$-\frac{256}{37367}$	−98307	98304	260
261	15	−19	31	−5	$-\frac{9}{4782976}$	$\frac{26811}{434816}$	27	$-\frac{2304}{37367}$	−884763	884736	261
262	23	−13	23	−1	$-\frac{81}{4782976}$	$\frac{241299}{434816}$	243	$-\frac{20736}{37367}$	−7962867	7962624	262
263	31	−17	15	3	$-\frac{729}{4782976}$	$\frac{2171691}{434816}$	2187	$-\frac{186624}{37367}$	−71665803	71663616	263

264	39	−21	7	7	$-\frac{6561}{4782976}$	$\frac{19545219}{434816}$	19683	$-\frac{1679616}{37367}$	−644992227	644972544	264
265	47	−25	−1	11	$-\frac{59049}{4782976}$	$\frac{175906971}{434816}$	177417	$-\frac{15116544}{37367}$	−5804930043	5804752896	265
266	55	−29	−9	15	$-\frac{531441}{4782976}$	$\frac{1583162739}{434816}$	1594323	$-\frac{136048896}{37367}$	−52244370387	52242776064	266
267	63	−33	−17	19	$-\frac{4782969}{4782976}$	$\frac{14248464651}{434816}$	14348907	$-\frac{1224440064}{37367}$	−470199333483	470184984576	267
268	71	−37	−25	23	$-\frac{43046721}{4782976}$	$\frac{128236181859}{434816}$	129140163	$-\frac{11019960576}{37367}$	−4231794001347	4231664861184	268
269	79	−41	−33	27	$-\frac{387420489}{4782976}$	$\frac{1154125636731}{434816}$	1162261467	$-\frac{99179645184}{37367}$	−38086146012123	38084983750656	269
270	−22	8	62	−16	$\frac{1}{31381105536}$	$-\frac{331}{316980864}$	$-\frac{1}{2187}$	$\frac{256}{245164887}$	$\frac{3641}{243}$	$-\frac{32768}{2187}$	270
271	−14	4	54	−12	$\frac{1}{3486789504}$	$-\frac{331}{35220096}$	$-\frac{1}{243}$	$\frac{256}{27240543}$	$\frac{3641}{27}$	$-\frac{32768}{243}$	271
272	−6	0	46	−8	$\frac{1}{387421056}$	$-\frac{331}{3913344}$	$-\frac{1}{27}$	$\frac{256}{3026727}$	$\frac{3641}{3}$	$-\frac{32768}{27}$	272
273	2	−4	38	−4	$\frac{1}{43046784}$	$-\frac{331}{434816}$	$-\frac{1}{3}$	$\frac{256}{336303}$	10923	$-\frac{32768}{3}$	273
274	10	−8	30	0	$\frac{1}{4782976}$	$-\frac{2979}{434816}$	−3	$\frac{256}{37367}$	98307	−98304	274
275	18	−12	22	4	$\frac{9}{4782976}$	$-\frac{26811}{434816}$	−27	$\frac{2304}{37367}$	884763	−884736	275
276	26	−16	14	8	$\frac{81}{4782976}$	$-\frac{241299}{434816}$	−243	$\frac{20736}{37367}$	7962867	−7962624	276
277	34	−20	6	12	$\frac{729}{4782976}$	$-\frac{2171691}{434816}$	−2187	$\frac{186624}{37367}$	71665803	−71663616	277
278	42	−24	−2	16	$\frac{6561}{4782976}$	$-\frac{19545219}{434816}$	−19683	$\frac{1679616}{37367}$	644992227	−644972544	278
279	50	−28	−10	20	$\frac{59049}{4782976}$	$-\frac{175906971}{434816}$	−177147	$\frac{15116544}{37367}$	5804930043	−5804752896	279
280	58	−32	−18	24	$\frac{531441}{4782976}$	$-\frac{1583162739}{434816}$	−1594323	$\frac{136048896}{37367}$	52244370387	−52242776064	280
281	66	−36	−26	28	$\frac{4782969}{4782976}$	$-\frac{14248464651}{434816}$	−14348907	$\frac{1224440064}{37367}$	470199333483	−470184984576	281
282	74	−40	−34	32	$\frac{43046721}{4782976}$	$-\frac{128236181859}{434816}$	−129140163	$\frac{11019960576}{37367}$	4231794001347	−4231664861184	282
283	−19	5	53	−7	$-\frac{1}{31381105536}$	$\frac{331}{316980864}$	$\frac{1}{2187}$	$-\frac{256}{245164887}$	$-\frac{3641}{243}$	$\frac{32768}{2187}$	283
284	−11	1	45	−3	$-\frac{1}{3486789504}$	$\frac{331}{35220096}$	$\frac{1}{243}$	$-\frac{256}{27240543}$	$-\frac{3641}{27}$	$\frac{32768}{243}$	284
285	−3	−3	37	1	$-\frac{1}{387421056}$	$\frac{331}{3913344}$	$\frac{1}{27}$	$-\frac{256}{3026727}$	$-\frac{3641}{3}$	$\frac{32768}{27}$	285

286	5	−7	29	5	$-\frac{1}{387421056}$	$\frac{331}{434816}$	$\frac{1}{3}$	$-\frac{256}{336303}$	−10923	$\frac{32768}{3}$	286
287	13	−11	21	9	$-\frac{1}{4782976}$	$\frac{2979}{434816}$	3	$-\frac{256}{37367}$	−98307	98304	287
288	21	−15	13	13	$\frac{9}{4782976}$	$-\frac{26811}{434816}$	−27	$\frac{2304}{37367}$	884763	−884736	288
289	29	−19	5	17	$-\frac{81}{4782976}$	$\frac{241299}{434816}$	243	$-\frac{20736}{37367}$	−7962867	7962624	289
290	37	−23	−3	21	$-\frac{729}{4782976}$	$\frac{2171691}{434816}$	2187	$-\frac{186624}{37367}$	−71665803	71663616	290
291	45	−27	−11	25	$-\frac{6561}{4782976}$	$\frac{19545219}{434816}$	19683	$-\frac{1679616}{37367}$	−644992227	644972544	291
292	53	−31	−19	29	$-\frac{59049}{4782976}$	$\frac{175906971}{434816}$	177147	$-\frac{15116544}{37367}$	−5804930043	5804752896	292
293	61	−35	−27	33	$-\frac{531441}{4782976}$	$\frac{1583162739}{434816}$	1594323	$-\frac{136048896}{37367}$	−52244370387	52242776064	293
294	69	−39	−35	37	$-\frac{4782969}{4782976}$	$\frac{14248464651}{434816}$	14348907	$-\frac{1224440064}{37367}$	−470199333483	470184984576	294
295	−16	2	44	2	$\frac{1}{31381105536}$	$-\frac{331}{316980864}$	$-\frac{1}{2187}$	$\frac{256}{245164887}$	$\frac{3641}{243}$	$-\frac{32768}{2187}$	295
296	−8	−2	36	6	$\frac{1}{3486789504}$	$-\frac{331}{35220096}$	$-\frac{1}{243}$	$\frac{256}{27240543}$	$\frac{3641}{27}$	$-\frac{32768}{243}$	296
297	0	−6	28	10	$\frac{1}{387421056}$	$-\frac{331}{3913344}$	$-\frac{1}{27}$	$\frac{256}{3026727}$	$\frac{3641}{3}$	$-\frac{32768}{27}$	297
298	8	−10	20	14	$\frac{1}{43046784}$	$-\frac{331}{434816}$	$-\frac{1}{3}$	$\frac{256}{336303}$	10923	$-\frac{32768}{3}$	298
299	16	−14	12	18	$\frac{1}{4782976}$	$-\frac{2979}{434816}$	−3	$\frac{256}{37367}$	98307	−98304	299
300	24	−18	4	22	$\frac{9}{4782976}$	$-\frac{26811}{434816}$	−27	$\frac{2304}{37367}$	884763	−884736	300
301	32	−22	−4	26	$\frac{81}{4782976}$	$-\frac{241299}{434816}$	−243	$\frac{20736}{37367}$	7962867	−7962624	301
302	40	−26	−12	30	$\frac{729}{4782976}$	$-\frac{2171691}{434816}$	−2187	$\frac{186624}{37367}$	71665803	−71663616	302
303	48	−30	−20	34	$\frac{6561}{4782976}$	$-\frac{19545219}{434816}$	−19683	$\frac{1679616}{37367}$	644992227	−644972544	303
304	56	−34	−28	38	$\frac{59049}{4782976}$	$-\frac{175906971}{434816}$	−177147	$\frac{15116544}{37367}$	5804930043	−5804752896	304
305	64	−38	−36	42	$\frac{531441}{4782976}$	$-\frac{1583162739}{434816}$	−1594323	$\frac{136048896}{37367}$	52244370387	−52242776064	305
306	−13	−1	35	11	$-\frac{1}{31381105536}$	$\frac{331}{316980864}$	$\frac{1}{2187}$	$-\frac{256}{245164887}$	$-\frac{3641}{243}$	$\frac{32768}{2187}$	306
307	−5	−5	27	15	$-\frac{1}{3486789504}$	$\frac{331}{35220096}$	$\frac{1}{243}$	$-\frac{256}{27240543}$	$-\frac{3641}{27}$	$\frac{32768}{243}$	307

308	3	−9	19	19	$-\frac{1}{387421056}$	$\frac{331}{3913344}$	$\frac{1}{27}$	$-\frac{256}{3026727}$	$-\frac{3641}{3}$	$\frac{32768}{27}$	308
309	11	−13	11	23	$-\frac{1}{43046784}$	$\frac{331}{434816}$	$\frac{1}{3}$	$-\frac{256}{336303}$	−10923	$\frac{32768}{3}$	309
310	19	−17	3	27	$-\frac{1}{4782976}$	$\frac{2979}{434816}$	3	$-\frac{256}{37367}$	−98307	98304	310
311	27	−21	−5	31	$-\frac{9}{4782976}$	$\frac{26811}{434816}$	27	$-\frac{2304}{37367}$	−884763	884736	311
312	35	−25	−13	35	$-\frac{81}{4782976}$	$\frac{241299}{434816}$	243	$-\frac{20736}{37367}$	−7962867	7962624	312
313	43	−29	−21	39	$-\frac{729}{4782976}$	$\frac{2171691}{434816}$	2187	$-\frac{186624}{37367}$	−71665803	71663616	313
314	51	−33	−29	43	$-\frac{6561}{4782976}$	$\frac{19545219}{434816}$	19683	$-\frac{1679616}{37367}$	−644992227	644972544	314
315	59	−37	−37	47	$-\frac{59049}{4782976}$	$\frac{175906971}{434816}$	177147	$-\frac{15116544}{37367}$	−5804930043	5804752896	315
316	−10	−4	26	20	$\frac{1}{31381105536}$	$-\frac{331}{316980864}$	$-\frac{1}{2187}$	$\frac{256}{245164887}$	$\frac{3641}{243}$	$-\frac{32768}{2187}$	316
317	−2	−8	18	24	$\frac{1}{3486789504}$	$-\frac{331}{35220096}$	$-\frac{1}{243}$	$\frac{256}{27240543}$	$\frac{3641}{27}$	$-\frac{32768}{243}$	317
318	6	−12	10	28	$\frac{1}{387421056}$	$-\frac{331}{3913344}$	$-\frac{1}{27}$	$\frac{256}{3026727}$	$\frac{3641}{3}$	$-\frac{32768}{27}$	318
319	14	−16	2	32	$\frac{1}{43046784}$	$-\frac{331}{434816}$	$-\frac{1}{3}$	$\frac{256}{336303}$	10923	$-\frac{32768}{3}$	319
320	22	−20	−6	36	$\frac{1}{4782976}$	$-\frac{2979}{434816}$	−3	$\frac{256}{37367}$	98307	−98304	320
321	30	−24	−14	40	$\frac{9}{4782976}$	$-\frac{26811}{434816}$	−27	$\frac{2304}{37367}$	884763	−884736	321
322	35	−28	−22	44	$\frac{81}{4782976}$	$-\frac{241299}{434816}$	−243	$\frac{20736}{37367}$	7962867	−7962624	322
323	46	−32	−30	48	$\frac{729}{4782976}$	$-\frac{2171691}{434816}$	−2187	$\frac{186624}{37367}$	71665803	−71663616	323
324	54	−36	−38	52	$\frac{6561}{4782976}$	$-\frac{19545219}{434816}$	−19683	$\frac{1679616}{37367}$	644992227	−644972544	324
325	−7	−7	17	29	$-\frac{1}{31381105536}$	$\frac{331}{316980864}$	$\frac{1}{2187}$	$-\frac{256}{245164887}$	$-\frac{3641}{243}$	$\frac{32768}{2187}$	325
326	1	−11	9	33	$-\frac{1}{3486789504}$	$\frac{331}{35220096}$	$\frac{1}{243}$	$-\frac{256}{27240543}$	$-\frac{3641}{27}$	$\frac{32768}{243}$	326
327	9	−15	1	37	$-\frac{1}{387421056}$	$\frac{331}{3913344}$	$\frac{1}{27}$	$-\frac{256}{3026727}$	$-\frac{3641}{3}$	$\frac{32768}{27}$	327
328	17	−19	−7	41	$-\frac{1}{43046784}$	$\frac{331}{434816}$	$\frac{1}{3}$	$-\frac{256}{336303}$	−10923	$\frac{32768}{3}$	328
329	25	−23	−15	45	$-\frac{1}{4782976}$	$\frac{2979}{434816}$	3	$-\frac{256}{37367}$	−98307	98304	329

330	33	−27	−23	49	$-\frac{9}{4782976}$	$\frac{26811}{434816}$	27	$-\frac{2304}{37367}$	−884763	884736	330
331	41	−31	−31	53	$-\frac{81}{4782976}$	$\frac{241299}{434816}$	243	$-\frac{20736}{37367}$	−7962867	7962624	331
332	49	−35	−39	57	$-\frac{729}{4782976}$	$\frac{2171691}{434816}$	2187	$-\frac{186624}{37367}$	−71665803	71663616	332
333	−4	−10	8	38	$\frac{1}{31381105536}$	$-\frac{331}{316980864}$	$-\frac{1}{2187}$	$\frac{256}{245164887}$	$\frac{3641}{243}$	$-\frac{32768}{2187}$	333
334	4	−14	0	42	$\frac{1}{3486789504}$	$-\frac{331}{35220096}$	$-\frac{1}{243}$	$\frac{256}{27240543}$	$\frac{3641}{27}$	$-\frac{32768}{243}$	334
335	12	−18	−8	46	$\frac{1}{387421056}$	$-\frac{331}{3913344}$	$-\frac{1}{27}$	$\frac{256}{3026727}$	$\frac{3641}{3}$	$-\frac{32768}{27}$	335
336	20	−22	−16	50	$\frac{1}{43046784}$	$-\frac{331}{434816}$	$-\frac{1}{3}$	$\frac{256}{336303}$	10923	$-\frac{32768}{3}$	336
337	28	−26	−24	54	$\frac{1}{4782976}$	$-\frac{2979}{434816}$	−3	$\frac{256}{37367}$	98307	−98304	337
338	36	−30	−32	58	$\frac{9}{4782976}$	$-\frac{26811}{434816}$	−27	$\frac{2304}{37367}$	884763	−884736	338
339	44	−34	−40	62	$\frac{81}{4782976}$	$-\frac{241299}{434816}$	−243	$\frac{20736}{37367}$	7962867	−7962624	339
340	−1	−13	−1	47	$-\frac{1}{31381105536}$	$\frac{331}{316980864}$	$\frac{1}{2187}$	$-\frac{256}{245164887}$	$-\frac{3641}{243}$	$\frac{32768}{2187}$	340
341	7	−17	−9	51	$-\frac{1}{3486789504}$	$\frac{331}{35220096}$	$\frac{1}{243}$	$-\frac{256}{27240543}$	$-\frac{3641}{27}$	$\frac{32768}{243}$	341
342	15	−21	−17	55	$-\frac{1}{387421056}$	$\frac{331}{3913344}$	$\frac{1}{27}$	$-\frac{256}{3026727}$	$-\frac{3641}{3}$	$\frac{32768}{27}$	342
343	23	−25	−25	59	$-\frac{1}{43046784}$	$\frac{331}{434816}$	$\frac{1}{3}$	$-\frac{256}{336303}$	−10923	$\frac{32768}{3}$	343
344	31	−29	−33	63	$-\frac{1}{4782976}$	$\frac{2979}{434816}$	3	$-\frac{256}{37367}$	−98307	98304	344
345	39	−33	−41	67	$-\frac{9}{4782976}$	$\frac{26811}{434816}$	27	$-\frac{2304}{37367}$	−884763	884736	345
346	2	−16	−10	56	$\frac{1}{31381105536}$	$-\frac{331}{316980864}$	$-\frac{1}{2187}$	$\frac{256}{245164887}$	$\frac{3641}{243}$	$-\frac{32768}{2187}$	346
347	10	−20	−18	60	$\frac{1}{3486789504}$	$-\frac{331}{35220096}$	$-\frac{1}{243}$	$\frac{256}{27240543}$	$\frac{3641}{27}$	$-\frac{32768}{243}$	347

348	18	−24	−26	64	$\frac{1}{387421056}$	$-\frac{331}{3913344}$	$-\frac{1}{27}$	$\frac{256}{3026727}$	$\frac{3641}{3}$	$-\frac{32768}{27}$	348
349	26	−28	−34	68	$\frac{1}{43046784}$	$-\frac{331}{434816}$	$-\frac{1}{3}$	$\frac{256}{336303}$	10923	$-\frac{32768}{3}$	349
350	34	−32	−42	72	$\frac{1}{4782976}$	$-\frac{2979}{434816}$	−3	$\frac{256}{37367}$	98307	−98304	350
351	5	−19	−19	65	$-\frac{1}{31381105536}$	$\frac{331}{316980864}$	$\frac{1}{2187}$	$-\frac{256}{245164887}$	$-\frac{3641}{243}$	$\frac{32768}{2187}$	351
352	13	−23	−27	69	$-\frac{1}{3486789504}$	$\frac{331}{35220096}$	$\frac{1}{243}$	$-\frac{256}{27240543}$	$-\frac{3641}{27}$	$\frac{32768}{243}$	352
353	21	−27	−35	73	$-\frac{1}{387421056}$	$\frac{331}{3913344}$	$\frac{1}{27}$	$-\frac{256}{3026727}$	$-\frac{3641}{3}$	$\frac{32768}{27}$	353
354	29	−31	−43	77	$-\frac{1}{43046784}$	$\frac{331}{434816}$	$\frac{1}{3}$	$-\frac{256}{336303}$	−10923	$\frac{32768}{3}$	354
355	8	−22	−28	74	$\frac{1}{31381105536}$	$-\frac{331}{316980864}$	$-\frac{1}{2187}$	$\frac{256}{245164887}$	$\frac{3641}{243}$	$-\frac{32768}{2187}$	355
356	16	−26	−36	78	$\frac{1}{3486789504}$	$-\frac{331}{35220096}$	$-\frac{1}{243}$	$\frac{256}{27240543}$	$\frac{3641}{27}$	$-\frac{32768}{243}$	356
357	24	−30	−44	82	$\frac{1}{387421056}$	$-\frac{331}{3913344}$	$-\frac{1}{27}$	$\frac{256}{3026727}$	$\frac{3641}{3}$	$-\frac{32768}{27}$	357
358	11	−25	−37	83	$-\frac{1}{31381105536}$	$\frac{331}{316980864}$	$\frac{1}{2187}$	$-\frac{256}{245164887}$	$-\frac{3641}{243}$	$\frac{32768}{2187}$	358
359	19	−29	−45	87	$-\frac{1}{3486789504}$	$\frac{331}{35220096}$	$\frac{1}{243}$	$-\frac{256}{27240543}$	$-\frac{3641}{27}$	$\frac{32768}{243}$	359
360	14	−28	−46	92	$\frac{1}{31381105536}$	$-\frac{331}{316980864}$	$-\frac{1}{2187}$	$\frac{256}{245164887}$	$\frac{3641}{243}$	$-\frac{32768}{2187}$	360

Remark 4 $S_{16}(\Gamma_0(12))$ is 27 dimensional, $M_{16}(\Gamma_0(12))$ is 33 dimensional, see [18] (Chapter 3, p. 87 and Chapter 5, p. 197), and generated by

$$\Delta_{1,16}, \Delta_{1,16}(2z), \Delta_{1,16}(3z), \Delta_{1,16}(4z), \Delta_{1,16}(6z), \Delta_{1,16}(12z),$$
$$\Delta_{2,16}, \Delta_{2,16}(2z), \Delta_{2,16}(3z), \Delta_{2,16}(6z),$$
$$\Delta_{3,16,1}, \Delta_{3,16,1}(2z), \Delta_{3,16,1}(4z), \Delta_{3,16,2}(z), \Delta_{3,16,2}(2z), \Delta_{3,16,2}(4z),$$
$$\Delta_{4,16}, \Delta_{4,16}(3z),$$
$$\Delta_{6,16,1}, \Delta_{6,16,1}(2z), \Delta_{6,16,2}, \Delta_{6,16,2}(2z), \Delta_{6,16,3}, \Delta_{6,16,3}(2z),$$
$$\Delta_{12,16,1}, \Delta_{12,16,2}, \Delta_{12,16,3}\left(\text{conjugate of } \Delta_{12,16,2}\right)$$

where $\Delta_{1,16}$ is the unique newform in $S_{16}(\Gamma_0(1))$; $\Delta_{2,16}$ is the unique newform in $S_{16}(\Gamma_0(2))$; $\Delta_{3,16,1}, \Delta_{3,16,2}$ are the unique newforms in $S_{16}(\Gamma_0(3))$, $\Delta_{4,16}$ is the unique newform in $S_{16}(\Gamma_0(4))$, $\Delta_{6,14,1}, \Delta_{6,14,2}, \Delta_{6,14,3}$ are the unique newforms in $S_{14}(\Gamma_0(6))$ and $\Delta_{12,16,1}, \Delta_{12,16,2}, \Delta_{12,16,3}$ are the unique newforms in $S_{16}(\Gamma_0(12))$. By taking t as a root of $x^2 - 69660x - 65283075900$, we see $f_1, \cdots, f_{27}$ as linear combinations inTable 3.

Table 3. Expression of f_i in terms of newforms.

$$\begin{aligned} f_1 = {} & \frac{29}{2211840000}\Delta_{1,16}(z) - \frac{29}{10240000}\Delta_{1,16}(2z) + \frac{17997}{81920000}\Delta_{1,16}(3z) + \frac{29}{67500}\Delta_{1,16}(4z) \\ & - \frac{485919}{10240000}\Delta_{1,16}(6z) + \frac{17997}{2500}\Delta_{1,16}(12z) - \frac{97}{4992000000}\Delta_{2,16}(z) - \frac{97}{39000000}\Delta_{2,16}(2z) \\ & + \frac{297861}{1664000000}\Delta_{2,16}(3z) + \frac{297861}{13000000}\Delta_{2,16}(6z) - \frac{517}{23601217536}\Delta_{3,16,1}(z) - \frac{517}{327794688}\Delta_{3,16,1}(2z) \\ & - \frac{517}{720252}\Delta_{3,16,1}(4z) + \frac{1861}{96111360000}\Delta_{3,16,2}(z) + \frac{24193}{5339520000}\Delta_{3,16,2}(2z) + \frac{29776}{46929375}\Delta_{3,16,2}(4z) \\ & - \frac{11}{6523846656}\Delta_{4,16}(z) - \frac{598035}{2174615552}\Delta_{4,16}(3z) - \frac{77}{2723217408}\Delta_{6,16,1}(z) + \frac{77}{21275136}\Delta_{6,16,1}(2z) \\ & + \frac{1367}{59050819584}\Delta_{6,16,2}(z) - \frac{1367}{461334528}\Delta_{6,16,2}(2z) + \frac{1547}{110592000000}\Delta_{6,16,3}(z) + \frac{1547}{864000000}\Delta_{6,16,3}(2z) \\ & - \frac{3133}{50594512896}\Delta_{12,16,1}(z) + (10111t + 98584027470)\Delta_{12,16,2}(z) + (10111t + 99288359730)\Delta_{12,16,3}(z). \end{aligned}$$

$$\begin{aligned} f_2 = {} & \frac{6341}{2972712960000}\Delta_{1,16}(z) - \frac{6341}{13762560000}\Delta_{1,16}(2z) + \frac{1148571}{36700160000}\Delta_{1,16}(3z) \\ & + \frac{6341}{90720000}\Delta_{1,16}(4z) - \frac{31011417}{4587520000}\Delta_{1,16}(6z) + \frac{1148571}{1120000}\Delta_{1,16}(12z) \\ & - \frac{4643}{1916928000000}\Delta_{2,16}(z) - \frac{4643}{14976000000}\Delta_{2,16}(2z) + \frac{5019003}{212992000000}\Delta_{2,16}(3z) \\ & + \frac{5019003}{1664000000}\Delta_{2,16}(6z) - \frac{4253}{1510477922304}\Delta_{3,16,1}(z) - \frac{4253}{20978860032}\Delta_{3,16,1}(2z) \\ & - \frac{4253}{46096128}\Delta_{3,16,1}(4z) + \frac{59381}{24604508160000}\Delta_{3,16,2}(z) + \frac{771953}{1366917120000}\Delta_{3,16,2}(2z) \\ & + \frac{59381}{750870000}\Delta_{3,16,2}(4z) - \frac{797}{1252578557952}\Delta_{4,16}(z) - \frac{5327451}{139175395328}\Delta_{4,16}(3z) \\ & - \frac{157}{33888927744}\Delta_{6,16,1}(z) + \frac{157}{264757248}\Delta_{6,16,1}(2z) + \frac{191}{52489617408}\Delta_{6,16,2}(z) \\ & - \frac{191}{410075136}\Delta_{6,16,2}(2z) + \frac{497}{294912000000}\Delta_{6,16,3}(z) + \frac{497}{2304000000}\Delta_{6,16,3}(2z) \\ & - \frac{1}{117116928}\Delta_{12,16,1}(z) + \frac{5599t + 33630313230}{737355880991577600}\Delta_{12,16,2}(z) \\ & + \frac{-5599t + 34020339570}{737355880991577600}\Delta_{12,16,3}(z). \end{aligned}$$

$$\begin{aligned} f_3 = {} & \frac{899}{495452160000}\Delta_{1,16}(z) - \frac{899}{2293760000}\Delta_{1,16}(2z) + \frac{512007}{18350080000}\Delta_{1,16}(3z) \\ & + \frac{899}{15120000}\Delta_{1,16}(4z) - \frac{13824189}{2293760000}\Delta_{1,16}(6z) + \frac{512007}{560000}\Delta_{1,16}(12z) \\ & - \frac{677}{319488000000}\Delta_{2,16}(z) - \frac{677}{2496000000}\Delta_{2,16}(2z) + \frac{2228151}{106496000000}\Delta_{2,16}(3z) \\ & + \frac{2228151}{832000000}\Delta_{2,16}(6z) - \frac{217}{83915440128}\Delta_{3,16,1}(z) - \frac{217}{1165492224}\Delta_{3,16,1}(2z) \\ & - \frac{217}{2560896}\Delta_{3,16,1}(4z) + \frac{2953}{1366917120000}\Delta_{3,16,2}(z) + \frac{38389}{75939840000}\Delta_{3,16,2}(2z) \\ & + \frac{2953}{41715000}\Delta_{3,16,2}(4z) - \frac{107}{208763092992}\Delta_{4,16}(z) - \frac{2396487}{69587697664}\Delta_{4,16}(3z) \\ & - \frac{611}{152500174848}\Delta_{6,16,1}(z) + \frac{611}{1191407616}\Delta_{6,16,1}(2z) + \frac{95}{29525409792}\Delta_{6,16,2}(z) \\ & - \frac{95}{230667264}\Delta_{6,16,2}(2z) + \frac{7}{4608000000}\Delta_{6,16,3}(z) + \frac{7}{36000000}\Delta_{6,16,3}(2z) \\ & - \frac{1535}{202378051584}\Delta_{12,16,1}(z) + \frac{10949t + 80222205690}{1990860878697725595520}\Delta_{12,16,2}(z) \\ & + \frac{-10949t + 80984913030}{1990860878697725595520}\Delta_{12,16,3}(z). \end{aligned}$$

$$
\begin{aligned}
f_4 = {} & \frac{1}{645120000}\Delta(z) - \frac{3}{8960000}\Delta(2z) + \frac{3533}{1143360000}\Delta(3z) + \frac{2}{39375}\Delta(4z) \\
& - \frac{95391}{17920000}\Delta(6z) - \frac{3533}{4375}\Delta(12z) - \frac{7}{3744000000}\Delta(z) - \frac{7}{29250000}\Delta(2z) \\
& + \frac{15419}{832000000}\Delta(3z) + \frac{15419}{6500000}\Delta(6z) - \frac{7}{2950152192}\Delta(z) - \frac{7}{40974336}\Delta(2z) \\
& - \frac{14}{180063}\Delta(4z) + \frac{47}{24027840000}\Delta(z) + \frac{611}{1334880000}\Delta(2z) + \frac{3008}{46929375}\Delta(4z) \\
& - \frac{1}{2446442496}\Delta(z) - \frac{16637}{543653888}\Delta(3z) - \frac{199}{57187565568}\Delta(z) + \frac{199}{446777856}\Delta(2z) \\
& + \frac{505}{177152458752}\Delta(z) - \frac{505}{1384003584}\Delta(2z) + \frac{151}{110592000000}\Delta(z) \\
& + \frac{151}{864000000}\Delta(2z) - \frac{1027}{151783538688}\Delta(z) + \frac{1613t + 16683315210}{4666080184447795200}\Delta(z) + \frac{-1613t + 16795676790}{4666080184447795200}\Delta(z).
\end{aligned}
$$

$$
\begin{aligned}
f_5 = {} & \frac{187}{139345920000}\Delta(z) - \frac{187}{645120000}\Delta(2z) + \frac{112591}{5160960000}\Delta(3z) \\
& + \frac{187}{4252500}\Delta(4z) - \frac{337773}{71680000}\Delta(6z) + \frac{112591}{157500}\Delta(12z) \\
& - \frac{151}{89856000000}\Delta(z) - \frac{151}{702000000}\Delta(2z) + \frac{496013}{29952000000}\Delta(3z) \\
& + \frac{496013}{234000000}\Delta(6z) - \frac{151}{70803652608}\Delta(z) - \frac{151}{983384064}\Delta(2z) \\
& - \frac{151}{2160756}\Delta(4z) + \frac{523}{288334080000}\Delta(z) + \frac{6799}{16018560000}\Delta(2z) \\
& + \frac{8368}{140788125}\Delta(4z) - \frac{19}{58714619904}\Delta(z) - \frac{532223}{19571539968}\Delta(3z) \\
& - \frac{29}{9531260928}\Delta(z) + \frac{29}{74462976}\Delta(2z) + \frac{37}{14762704896}\Delta(z) \\
& - \frac{37}{11533632}\Delta(2z) + \frac{11}{9216000000}\Delta(z) + \frac{11}{72000000}\Delta(2z) \\
& - \frac{229}{37945884672}\Delta(z) + \frac{811t + 29641066470}{9332160368895590400}\Delta(z) \\
& + \frac{-811t + 29697560730}{9332160368895590400}\Delta(z).
\end{aligned}
$$

$$
\begin{aligned}
f_6 = {} & -\frac{527}{46448640000}\Delta(z) + \frac{527}{215040000}\Delta(2z) - \frac{61137}{573440000}\Delta(3z) \\
& - \frac{527}{1417500}\Delta(4z) + \frac{1650699}{1680000}\Delta(6z) - \frac{61137}{17500}\Delta(12z) \\
& - \frac{629}{29952000000}\Delta(z) - \frac{629}{234000000}\Delta(2z) + \frac{587709}{3328000000}\Delta(3z) \\
& + \frac{587709}{26000000}\Delta(6z) - \frac{5}{23601217536}\Delta(z) - \frac{5}{327794688}\Delta(2z) \\
& - \frac{5}{720252}\Delta(4z) + \frac{1217}{96111360000}\Delta(z) + \frac{15821}{5339520000}\Delta(2z) \\
& + \frac{19472}{46929375}\Delta(4z) + \frac{15}{2174615552}\Delta(z) + \frac{3105}{2174615552}\Delta(3z) \\
& + \frac{23}{1059028992}\Delta(z) - \frac{23}{8273664}\Delta(2z) - \frac{25}{1640300544}\Delta(z) \\
& + \frac{25}{12814848}\Delta(2z) + \frac{371}{27648000000}\Delta(z) + \frac{371}{216000000}\Delta(2z) \\
& - \frac{5}{1405403136}\Delta(z) + \frac{2249t - 655556670}{345635569218355200}\Delta(z) \\
& - \frac{2249t + 498891330}{345635569218355200}\Delta(z).
\end{aligned}
$$

$$
\begin{aligned}
f_7 = &-\frac{1218595428124173875327363153}{471400156632462524747284296499200000}\Delta_{1,16}(z)\\
&-\frac{95114013943327924473760149421031544657928255699670157627333095511}{614410784113929460711766883942751048355211990391340448642498560000}\Delta_{1,16}(2z)\\
&-\frac{213022258764287126157166537}{4827001675549232787016908800000}\Delta_{1,16}(3z)\\
&-\frac{37940907346454430218531589470622504218751469933947801511805763539379 9}{85383028086421079466561721371455844219952254547031871171641344000000}\Delta_{1,16}(4z)\\
&-\frac{264071136274916671784362679027533390235320933707342857992923950831121 21}{61441078411392946071176688394275104835521199039134044864249856000000}\Delta_{1,16}(6z)\\
&+\frac{1969424271653554293105803759254452282916421075920629041644919622330879781}{53760943609968827812279602344990716731081049159242289256218624000000}\Delta_{1,16}(12z)\\
&+\frac{36720058380953696452825273}{847332736090497242938802995200000 0}\Delta_{2,16}(z)\\
&+\frac{331966236813599436096274663078264332755623854057133014180288648984491 3}{179715154353324367258191813553254681643899507189467081227930828800000 00}\Delta_{2,16}(2z)\\
&-\frac{145634444765492327012883569}{3663349485907899883003904000000}\Delta_{2,16}(3z)\\
&-\frac{636030325665152420042349173492978574499766251879483863304866620825649967 3}{1996835048370270747313242372813940907154438968771836458088120320000000}\Delta_{2,16}(6z)\\
&+\frac{11864785475793655505685 5}{25979419509406884495097382043648}\Delta_{3,16,1}(z)\\
&+\frac{83242702806757313708276633133965682061060259117149650553416928792793}{19668026492427818752736512075268192359108362066815277369584749903872 0}\Delta_{3,16,1}(2z)\\
&-\frac{49756637743618857700249760580515455035474881961970915865991817257853 19}{22126529803981296096828576084676716403996907325167187040782843641856 0}\Delta_{3,16,1}(4z)\\
&-\frac{23062692139224271815795569}{3301398456967329570302282792960000}\Delta_{3,16,2}(z)\\
&-\frac{761384780650862586981988006208288587854490088549099976503213805246022 7}{128150729296562837267956777818359299910688340895873834075609137152000 00}\Delta_{3,16,2}(2z)\\
&-\frac{728172462823167914015591414301388299175766237158674678312623367650708191 9}{576678281834532767705805500182616849598097534031432253340241117184000 00}\Delta_{3,16,2}(4z)\\
&+\frac{16927175898389975 48293}{595889219416545173929134234009 6}\Delta_{4,16}(z)+\frac{18986052708980720014329483 1}{331049566342523096627296796672 0}\Delta_{4,16}(3z)\\
&+\frac{56965527997151011903936 31}{10229396431828960769944594179686 40}\Delta_{5,16,1}(z)\\
&+\frac{76348744890692672463532449515779992822503720200675195553124845005509999}{85782854009281332560012325743669731289341856091417402065804255349964 80}\Delta_{5,16,1}(2z)\\
&-\frac{18148018440531592480125 1}{39610068890164772906176436109312}\Delta_{5,16,2}(z)\\
&-\frac{88332540284712745105418185210418212158613084627116337754328435011033}{33216669033669087419854396810518730536850417960210497792397888349798 4}\Delta_{5,16,2}(2z)\\
&-\frac{3834283475428835803619 9}{12363804514939162105138176000000}\Delta_{5,16,3}(z)\\
&-\frac{626127467479397875509947117424954778205978983246274983122313591867 79}{129602274774031995618888327081674049262427529223173375885527040000000}\Delta_{5,16,3}(2z)\\
&+\frac{351527729085893253562078 1}{27150202559666339487146178537062 4}\Delta_{12,16,1}(z)\\
&+\frac{-4193821676956890858762 67r-37475783522213133243331979970 30}{5682673685140249072091908264204802457 60}\Delta_{12,16,2}(z)\\
&+\frac{4193821676956890858762 67r-37767925140229950260553387562 50}{5682673685140249072091908264204802457 60}\Delta_{12,16,3}(z)\\
&+\frac{20776784213229 3}{12804139123145291671075245260800}E_{16}(z)\\
&+\frac{3893526571326585547765605380590304837054861224690877 73}{72193691802162018469035367894849100235268204566303480971264 00}E_{16}(2z)\\
&-\frac{2343816}{190054142030133966098225}E_{16}(3z)\\
&-\frac{1272830915791951666788081818704016661173962382448234062558 1}{1443873836043240369380707357896982004705364091326069619425280 0}E_{16}(4z)\\
&-\frac{6362803276892031175411753128029842826781969702641867352 9}{61181094747594930905962176182075508673956105564663966924800}E_{16}(6z)\\
&+\frac{6935936546517597174149866024128733438060610717383151483691}{36096845901081009234517683947424550117634102283151740485632 00}E_{16}(12z)
\end{aligned}
$$

$$f = \frac{112591}{1720320000}\Delta(z) - \frac{1013319}{71680000}\Delta(2z) + \frac{99379467}{573440000}\Delta(3z) + \frac{112591}{52500}\Delta(4z) - \frac{2683245609}{71680000}\Delta(6z) + \frac{99379467}{17500}\Delta(12z)$$
$$+\frac{496013}{9984000000}\Delta(z) + \frac{496013}{78000000}\Delta(2z) - \frac{722228319}{3328000000}\Delta(3z) - \frac{722228319}{26000000}\Delta(6z) + \frac{1359}{32374784}\Delta(z) + \frac{12231}{4046848}\Delta(2z)$$
$$+\frac{1359}{988}\Delta(4z) + \frac{4707}{131840000}\Delta(z) + \frac{550719}{65920000}\Delta(2z) + \frac{75312}{64375}\Delta(4z) - \frac{532223}{6523846656}\Delta(z) - \frac{90876411}{2174615552}\Delta(3z)$$
$$+\frac{261}{4358144}\Delta(z) - \frac{261}{34048}\Delta(2z) + \frac{333}{6750208}\Delta(z) - \frac{333}{52736}\Delta(2z) + \frac{24057}{1024000000}\Delta(z) + \frac{24057}{8000000}\Delta(2z)$$
$$-\frac{687}{5783552}\Delta(z) + \frac{-81\ell - 29641066470}{474122865868800}\Delta(z) + \frac{81\ell - 29697560730}{474122865868800}\Delta(z).$$

$$f = \frac{690031126457668339015748339309}{65472243976730906214900596736000\,0}\Delta(z) - \frac{30313733589758327316374127526990319269014849828612891068249265959 6077}{130639092972224251583839433698327441656526949456958762892611256320000 0}\Delta(2z)$$
$$+\frac{799899248786273165894110437249}{603375209443654098377113600000}\Delta(3z) + \frac{31962553103687717441752839695825821595082965832332963724340268833270 3}{96038156359722371439988193654288782474744628636541085506809651200000}\Delta(4z)$$
$$-\frac{144452844689852200011973956319391343390078921158974824977270464465007161}{48384849248971945031051642110491645057972944243318060330596761600000}\Delta(6z)$$
$$+\frac{26902204061480287600842201157722882391538692804448514013047724734 61239443}{604810615612149312888145526381145563224661803041475754132459520000 0}\Delta(12z)$$
$$-\frac{1546936517995753160934757826897}{1529906329052286688639505408000000}\Delta(z) - \frac{753217210497202996664874202406887105693225974056389430994758972431988 3}{6065386459424697394963973707422345505481608367644513991442665472000000 0}\Delta(2z)$$
$$+\frac{6533460869849671919028469110396 9}{59529429146003373098813440000 00}\Delta(3z) + \frac{265751461970152598336548429342624293809893139995876487236020022500232771 29}{20217954864748991316546579024741151684938694558815046638142218240000000}\Delta(6z)$$
$$-\frac{6499511179122327803628975 9}{57910228673232081350525714432}\Delta(z) - \frac{45667778306690858421169710010000976479289998837345895690984544836003}{663795894119438882904857282540301492119907219755015611223485309255680}\Delta(2z)$$
$$-\frac{3104496502589097270819711314202655408789815930753729362125470131197739}{829744867649298603631071603175376865149884024693769514029356636 56960}\Delta(4z) + \frac{1403546857583570285578054947}{94331249262128421987350528 0000}\Delta(z)$$
$$+\frac{1432928592763864354362958109186168987268502990481442251373425724603769 7}{432508711375899575779354125136962637198573150523574190005180837888 00000}\Delta(2z)$$
$$+\frac{976862495070941011542240911770100305605653284833136704510347782849426781}{216254355687949787889677062568481318599286575261787095002590418944000 00}\Delta(4z)$$
$$-\frac{40004151721036771462802432 93}{116694972135740095561122120826 88}\Delta(z) + \frac{159030405291697987047886934497899}{1944916202262334942685368680 4480}\Delta(3z)$$
$$-\frac{649061581281262957406279 7}{4872254816257506844395192320}\Delta(z) + \frac{699750625725553947633992259869926144901086892231804018604718993470651}{3574285583720055523333846905986238803722577337142391752741843972915 20}\Delta(2z)$$
$$+\frac{174112047536898594423933 81}{120743999055524380143808675 84}\Delta(z) - \frac{265901836267556255239023286219804735178554111149460451081037486065 3}{13840278764028786424939332004382804390354340816754374080165786812416}\Delta(2z)$$
$$+\frac{2052975379400861685204768993}{18316747429539499415019520000 00}\Delta(z) + \frac{189185955502464839917980520259753135726141478214715544177391439647 63}{145802559120785995071249367966883305420230970376070047871217920000000}\Delta(2z)$$
$$+\frac{1322559797897159327281773 9}{10345298948203909269603024896}\Delta(z) + \frac{-141909938796823710450122633r - 108679994371740298163715735293421 0}{84808484225471125586902304151306240 0}\Delta(z)$$
$$+\frac{141909938796823710450122633\ell - 109668539005398972130711289554899 0}{84808484225471125586902304151306240 0}\Delta(z)$$
$$+\frac{2211028162513}{480155217117948437665321697280 0}E(z) + \frac{12430263719850791753713866960057397717948624552591579}{81217903277432270777664788881705237764676730137091416092672 00}E(2z)$$
$$-\frac{269049582472896}{19005414203013396609822 5}E(3z) - \frac{121907147653155578656075807739511341447291233400059212988 9}{487307419664593624665988733290231426588060380822548496556032 00}E(4z)$$
$$-\frac{1233051489489362034181672002119412749555317836620300060 7}{688287315910442972692074482048349472582006187602469627904 00}E(6z)$$
$$+\frac{523016583759503354625382815866396720047661541839417263099 9}{1218268549161484061664971833225578566470150952056371241390080 0}E(12z)$$

$$f = \frac{53}{725760000}\Delta(z) - \frac{53}{3360000}\Delta(2z) + \frac{243}{8960000}\Delta(3z) + \frac{848}{354375}\Delta(4z) - \frac{6561}{1120000}\Delta(6z)$$
$$+\frac{3888}{4375}\Delta(12z) + \frac{659}{29952000000}\Delta(z) + \frac{659}{234000000}\Delta(2z) + \frac{17661}{3328000000}\Delta(3z) + \frac{176661}{26000000}\Delta(6z)$$
$$-\frac{1}{40974336}\Delta(z) - \frac{1}{569088}\Delta(2z) - \frac{16}{20007}\Delta(4z) - \frac{7}{166860000}\Delta(z) - \frac{91}{9270000}\Delta(2z)$$
$$-\frac{7168}{5214375}\Delta(4z) - \frac{27}{271826944}\Delta(z) - \frac{5589}{271826944}\Delta(3z) - \frac{1}{58834944}\Delta(z) + \frac{1}{459648}\Delta(2z)$$
$$-\frac{1}{45563904}\Delta(z) + \frac{1}{355968}\Delta(2z) + \frac{71}{6912000000}\Delta(z) + \frac{71}{54000000}\Delta(2z) + \frac{1}{19519488}\Delta(z)$$
$$+\frac{-7t + 1783074}{64006586892288}\Delta(z) + \frac{7t + 1295454}{64006586892288}\Delta(z).$$

$$f = \frac{836863}{6720000}\Delta(z) - \frac{7531767}{280000}\Delta(2z) + \frac{686090331}{2240000}\Delta(3z) + \frac{53559232}{13125}\Delta(4z) - \frac{18524438937}{280000}\Delta(6z) + \frac{43909781184}{4375}\Delta(12z) + \frac{12840549}{104000000}\Delta(z) + \frac{12840549}{812500}\Delta(2z) - \frac{60595434261}{104000000}\Delta(3z) - \frac{60595434261}{812500}\Delta(6z) + \frac{11079}{126464}\Delta(z) + \frac{99711}{15808}\Delta(2z) + \frac{709056}{247}\Delta(4z) + \frac{43101}{515000}\Delta(z) + \frac{5042817}{257500}\Delta(2z) + \frac{176541696}{64375}\Delta(4z) + \frac{1062583}{12741888}\Delta(z) + \frac{52612659}{4247296}\Delta(3z) + \frac{6855}{68096}\Delta(z) - \frac{6855}{532}\Delta(2z) + \frac{17601}{210944}\Delta(z) - \frac{17601}{1648}\Delta(2z) + \frac{2024433}{32000000}\Delta(z) + \frac{2024433}{250000}\Delta(2z) + \frac{1401}{11296}\Delta(z) + \frac{7193r + 58049091810}{926021222400}\Delta(z) + \frac{-7193r + 58550156190}{926021222400}\Delta(z).$$

$$f = -\frac{473}{120960000}\Delta(z) + \frac{473}{560000}\Delta(2z) - \frac{102789}{4480000}\Delta(3z) - \frac{15136}{118125}\Delta(4z) + \frac{2775303}{560000}\Delta(6z) - \frac{3289248}{4375}\Delta(12z) + \frac{1457}{624000000}\Delta(z) + \frac{1457}{4875000}\Delta(2z) + \frac{452709}{208000000}\Delta(3z) + \frac{452709}{1625000}\Delta(6z) - \frac{11}{6829056}\Delta(z) - \frac{11}{94848}\Delta(2z) - \frac{352}{6669}\Delta(4z) + \frac{37}{27810000}\Delta(z) + \frac{481}{1545000}\Delta(2z) + \frac{75776}{1738125}\Delta(4z) + \frac{1}{1225728}\Delta(z) - \frac{1}{9576}\Delta(2z) + \frac{5}{1898496}\Delta(z) - \frac{5}{14832}\Delta(2z) - \frac{51}{32000000}\Delta(z) - \frac{51}{250000}\Delta(2z).$$

$$f = -\frac{43}{17418240000}\Delta(z) + \frac{43}{80640000}\Delta(2z) + \frac{1601}{645120000}\Delta(3z) - \frac{86}{1063125}\Delta(4z) - \frac{4803}{8960000}\Delta(6z) + \frac{3202}{39375}\Delta(12z) - \frac{113}{89856000000}\Delta(z) - \frac{113}{702000000}\Delta(2z) - \frac{83381}{29952000000}\Delta(3z) - \frac{83381}{234000000}\Delta(6z) + \frac{7}{8850456576}\Delta(z) + \frac{7}{122923008}\Delta(2z) + \frac{14}{540189}\Delta(4z) + \frac{53}{36041760000}\Delta(z) + \frac{689}{2002320000}\Delta(2z) + \frac{6784}{140788125}\Delta(4z) - \frac{1}{611610624}\Delta(z) - \frac{23}{67956736}\Delta(3z) + \frac{5}{4765630464}\Delta(z) - \frac{5}{37231488}\Delta(2z) + \frac{1}{3690676224}\Delta(z) - \frac{1}{28833408}\Delta(2z) + \frac{1}{6912000000}\Delta(z) + \frac{1}{54000000}\Delta(2z) + \frac{1}{1185808896}\Delta(z) + \frac{-619r + 137004570}{29163001152798720 0}\Delta(z) + \frac{619r + 93885030}{291630011527987200}\Delta(z).$$

$$\begin{aligned}
f_{16} = &-\frac{5999}{737280000}\Delta_{1,36}(z)+\frac{17997}{10240000}\Delta_{1,36}(2z)-\frac{1712421}{81920000}\Delta_{1,36}(3z)\\
&-\frac{5999}{22500}\Delta_{1,36}(4z)+\frac{46235367}{10240000}\Delta_{1,36}(6z)-\frac{1712421}{2500}\Delta_{1,36}(12z)\\
&-\frac{99287}{14976000000}\Delta_{2,36}(z)-\frac{99287}{117000000}\Delta_{2,36}(2z)+\frac{51549777}{1664000000}\Delta_{2,36}(3z)\\
&+\frac{51549777}{13000000}\Delta_{2,36}(6z)-\frac{517}{97124352}\Delta_{3,36,1}(z)-\frac{1551}{4046848}\Delta_{3,36,1}(2z)\\
&-\frac{517}{2964}\Delta_{3,36,1}(4z)-\frac{1861}{395520000}\Delta_{3,36,2}(z)-\frac{7259}{65920000}\Delta_{3,36,2}(2z)\\
&-\frac{29776}{193125}\Delta_{3,36,2}(4z)+\frac{199345}{19571539968}\Delta_{4,36}(z)+\frac{5845851}{2174615552}\Delta_{4,36}(3z)\\
&-\frac{77}{11206656}\Delta_{6,36,1}(z)+\frac{77}{87552}\Delta_{6,36,1}(2z)-\frac{1367}{243007488}\Delta_{6,36,2}(z)\\
&+\frac{1367}{1898496}\Delta_{6,36,2}(2z)-\frac{13923}{4096000000}\Delta_{6,36,3}(z)-\frac{13923}{32000000}\Delta_{6,36,3}(2z)\\
&+\frac{3133}{208207872}\Delta_{12,36,1}(z)+\frac{10111r+98584027470}{12801317378457600}\Delta_{12,36,2}(z)\\
&+\frac{-10111r+99288359730}{12801317378457600}\Delta_{12,36,3}(z).
\end{aligned}$$

$$\begin{aligned}
f_{17} = &-\frac{419}{362880000}\Delta_{1,36}(2z)+\frac{21017}{45360000}\Delta_{1,36}(4z)-\frac{82989}{4480000}\Delta_{1,36}(6z)+\frac{4081527}{560000}\Delta_{1,36}(12z)\\
&-\frac{6229}{1872000000}\Delta_{2,36}(2z)+\frac{6495309}{208000000}\Delta_{2,36}(6z)-\frac{49}{1843845121}\Delta_{3,36,1}(2z)\\
&-\frac{19313}{23048064}\Delta_{3,36,1}(4z)+\frac{1249}{750870000}\Delta_{3,36,2}(2z)+\frac{295697}{375435000}\Delta_{3,36,2}(4z)\\
&+\frac{115}{33094656}\Delta_{6,36,1}(2z)-\frac{143}{51259392}\Delta_{6,36,2}(2z)+\frac{691}{288000000}\Delta_{6,36,3}(2z)
\end{aligned}$$

$$\begin{aligned}
f_{18} = &-\frac{181}{181440000}\Delta_{1,36}(2z)+\frac{1151}{2835000}\Delta_{1,36}(4z)-\frac{36411}{2240000}\Delta_{1,36}(6z)+\frac{227481}{35000}\Delta_{1,36}(12z)\\
&-\frac{2771}{936000000}\Delta_{2,36}(2z)+\frac{2873691}{104000000}\Delta_{2,36}(6z)-\frac{23}{92192256}\Delta_{3,36,1}(2z)\\
&-\frac{1067}{1440504}\Delta_{3,36,1}(4z)+\frac{551}{375435000}\Delta_{3,36,2}(2z)+\frac{32632}{46929375}\Delta_{3,36,2}(4z)\\
&+\frac{461}{148925952}\Delta_{6,36,1}(2z)-\frac{577}{230567264}\Delta_{6,36,2}(2z)+\frac{103}{48000000}\Delta_{6,36,3}(2z)
\end{aligned}$$

$$\begin{aligned}
f_{19} = &\frac{30164674243500778225439808498801255783}{5845798685924840125532301792721344000 0}\Delta_{1,36}(2z)+\frac{10822845902179562208713922146349462508 1}{73072483574060501569153772409016800 00}\Delta_{1,36}(4z)\\
&+\frac{310177214518494902505208935294294916741 9}{21651106244166074539008525158227200 00}\Delta_{1,36}(6z)-\frac{33047294429459075615400456620997496510267}{270638828052075931737606564477840000}\Delta_{1,36}(12z)\\
&-\frac{1299728638781507717519443312086174923 29}{2110982858806192267555333120292715200 0000}\Delta_{2,36}(2z)+\frac{74706945544725070254562828372084881714 0227}{70366095293539742251777706764238400 0000}\Delta_{2,36}(6z)\\
&-\frac{29332800951290575846259872152295739797}{20792336766097471358493291016351276 3392}\Delta_{3,36,1}(2z)+\frac{19481867408289411135083256727109427703 3}{25990420957621839198116613770439095424}\Delta_{3,36,1}(4z)\\
&+\frac{16768360546320261294314753211420987529 9}{84672740341442606193256933778948217000 0}\Delta_{3,36,2}(2z)+\frac{17818757149853870998254485587531747229597}{42336370170721303096628456889474108 5000}\Delta_{3,36,2}(4z)\\
&-\frac{14234744479773888084180581806762011023}{4798231561407108775036913311465679155 2}\Delta_{6,36,1}(2z)+\frac{46113624946307175246214068266935663717}{52022931665782337245137060113785784524 8}\Delta_{6,36,2}(2z)\\
&+\frac{17432668584289644612623510125778279197}{10825553122083037269504262579113600000 0}\Delta_{6,36,3}(2z)-\frac{17230868652997867242604 68}{95851251601776892490402324919235}E_{16}(2z)\\
&+\frac{2816467529601024962810692952 4}{95851251601776892490402324919235}E_{16}(4z)+\frac{66454520325448269758836005 72}{19170250320355378498080464983847}E_{16}(6z)\\
&-\frac{6139021237186908472080067191 6}{95851251601776892490402324919235}E_{16}(12z).
\end{aligned}$$

$$f_{18}=-\frac{53}{60480000}\Delta_{1,1,6}(2z)+\frac{169}{472500}\Delta_{1,1,6}(4z)-\frac{95987}{6720000}\Delta_{1,1,6}(6z)+\frac{304751}{52500}\Delta_{1,1,6}(12z)-\frac{823}{312000000}\Delta_{2,1,6}(2z)$$
$$+\frac{7629047}{312000000}\Delta_{2,1,6}(6z)-\frac{7}{30730752}\Delta_{3,2,6,3}(2z)-\frac{157}{240084}\Delta_{3,2,6,3}(4z)+\frac{163}{125145000}\Delta_{3,1,6,2}(2z)+\frac{9616}{15643125}\Delta_{3,1,6,2}(4z)$$
$$+\frac{137}{49641984}\Delta_{4,3,6,3}(2z)-\frac{43}{19222272}\Delta_{4,3,6,2}(2z)-\frac{23}{12000000}\Delta_{4,3,6,3}(2z)$$

$$f_{19}=-\frac{19}{15120000}\Delta_{1,1,6}(2z)-\frac{517}{1890000}\Delta_{1,1,6}(4z)+\frac{5967}{560000}\Delta_{1,1,6}(6z)+\frac{86319}{70000}\Delta_{1,1,6}(12z)$$
$$+\frac{29}{78000000}\Delta_{2,1,6}(2z)-\frac{240327}{26000000}\Delta_{2,1,6}(6z)-\frac{11}{7682688}\Delta_{3,2,6,3}(2z)+\frac{181}{960336}\Delta_{3,2,6,3}(4z)$$
$$+\frac{13}{15643125}\Delta_{3,1,6,2}(2z)-\frac{547}{15643125}\Delta_{3,1,6,2}(4z)+\frac{1}{1378944}\Delta_{4,3,6,2}(2z)$$
$$-\frac{11}{8543232}\Delta_{4,3,6,2}(2z)-\frac{67}{144000000}\Delta_{4,3,6,3}(2z)+\frac{1327}{1122729984}\Delta_{12,1,6,2}(z).$$

$$f_{20}=-\frac{1789863301316506331707525830936396010397}{409205908014738808787261125490494080000}\Delta_{1,1,6}(2z)$$
$$+\frac{204390035055363928988407427741855739284 71}{511507385018423510984076406863117 60000}\Delta_{1,1,6}(4z)$$
$$-\frac{4281182792322725937128144783764746 72266921}{151557743709162521773059676107590 40000}\Delta_{1,1,6}(6z)$$
$$+\frac{591324181296558869796917920301181 9458687803}{189447179636453152216324595134 4880000}\Delta_{1,1,6}(12z)$$
$$-\frac{5607431939844561001690326523666 4233705843}{189988457292557304079799808263 44368000000}\Delta_{2,1,6}(2z)$$
$$-\frac{23018447436087765658975155678061034153732609}{6332948576418576802659993608781456000000}\Delta_{2,1,6}(6z)$$
$$-\frac{1703613821347982006720984807029 46097331}{207923367660974713584932910163512763392}\Delta_{3,2,6,3}(2z)$$
$$-\frac{4415573731700296730438764143151963140211}{2599042095762183919811661377043909 5424}\Delta_{3,2,6,3}(4z)$$
$$-\frac{687246025785522461886346195449765332 98883}{13547638454630816990921109404631714720000}\Delta_{3,1,6,2}(2z)$$
$$-\frac{611869586300774214085769719509229406 0311199}{677381922731540849546055470231585736 0000}\Delta_{3,1,6,2}(4z)$$
$$+\frac{9379699084691278165991813355713323 3667}{111958736432832538084194643934199180288}\Delta_{4,3,6,3}(2z)$$
$$-\frac{784112211421316372012943841929523742 87}{216762215274093071854737750474107 43552}\Delta_{4,3,6,2}(2z)$$
$$-\frac{731234530042280556976123929402546347833}{182681208935151253922884431022542000000}\Delta_{4,3,6,3}(2z)$$
$$+\frac{21602130308836605557255 51077}{9201720153770581679078623192246560}E_{18}(2z)$$
$$-\frac{21182655306486843403636526345 8453}{5521032092262349007447173915347 9360}E_{18}(4z)$$
$$-\frac{587331790961553566161010528 3501}{2300430038442645419769655798061640}E_{18}(6z)$$
$$-\frac{70554644323491197590334687391203}{1104206418452469801489434783069587 2}E_{18}(12z).$$

$$f_{21}=-\frac{4437}{80000}\Delta_{1,1,6}(2z)+\frac{8874}{625}\Delta_{1,1,6}(4z)+\frac{14348907}{80000}\Delta_{1,1,6}(6z)+\frac{28697814}{625}\Delta_{1,1,6}(12z)$$
$$+\frac{2618849}{6500000}\Delta_{2,1,6}(2z)-\frac{11809150461}{6500000}\Delta_{2,1,6}(6z)+\frac{243}{1664}\Delta_{3,2,6,3}(2z)+\frac{486}{13}\Delta_{3,2,6,3}(4z)$$
$$+\frac{243}{1250}\Delta_{3,1,6,2}(2z)+\frac{31104}{625}\Delta_{3,1,6,2}(4z)+\frac{50301}{250000}\Delta_{4,3,6,3}(2z)$$

$$\begin{aligned}
f_{\text{[illegible]}} = {} & \frac{29}{2211840000}\Delta_{\text{[illegible]}}(z) - \frac{29}{10240000}\Delta_{\text{[illegible]}}(2z) + \frac{17997}{81920000}\Delta_{\text{[illegible]}}(3z) \\
& + \frac{29}{67500}\Delta_{\text{[illegible]}}(4z) - \frac{485919}{10240000}\Delta_{\text{[illegible]}}(6z) + \frac{17997}{2500}\Delta_{\text{[illegible]}}(12z) \\
& - \frac{97}{4992000000}\Delta_{\text{[illegible]}}(z) - \frac{97}{39000000}\Delta_{\text{[illegible]}}(2z) + \frac{297861}{1664000000}\Delta_{\text{[illegible]}}(3z) \\
& + \frac{297861}{13000000}\Delta_{\text{[illegible]}}(6z) - \frac{517}{23601217536}\Delta_{\text{[illegible]}}(z) - \frac{517}{327794688}\Delta_{\text{[illegible]}}(2z) \\
& - \frac{517}{720252}\Delta_{\text{[illegible]}}(4z) + \frac{1861}{96111360000}\Delta_{\text{[illegible]}}(z) + \frac{24193}{5339520000}\Delta_{\text{[illegible]}}(2z) \\
& + \frac{29776}{46929375}\Delta_{\text{[illegible]}}(4z) - \frac{11}{6523846656}\Delta_{\text{[illegible]}}(z) - \frac{598035}{2174615552}\Delta_{\text{[illegible]}}(3z) \\
& - \frac{77}{2723217408}\Delta_{\text{[illegible]}}(z) + \frac{77}{21275136}\Delta_{\text{[illegible]}}(2z) + \frac{1367}{59050819584}\Delta_{\text{[illegible]}}(z) \\
& - \frac{1367}{461334528}\Delta_{\text{[illegible]}}(2z) + \frac{1547}{110592000000}\Delta_{\text{[illegible]}}(z) \\
& + \frac{1547}{864000000}\Delta_{\text{[illegible]}}(2z) - \frac{3133}{50594512896}\Delta_{\text{[illegible]}}(z) \\
& + \frac{10111r + 9858027470}{311072012296519680 0}\Delta_{\text{[illegible]}}(z) \\
& + \frac{-10111r + 99288359730}{3110720122965196800}\Delta_{\text{[illegible]}}(z).
\end{aligned}$$

$$\begin{aligned}
f_{\text{[illegible]}} = {} & \frac{29}{2211840000}\Delta_{\text{[illegible]}}(z) - \frac{29}{10240000}\Delta_{\text{[illegible]}}(2z) + \frac{17997}{81920000}\Delta_{\text{[illegible]}}(3z) \\
& + \frac{29}{67500}\Delta_{\text{[illegible]}}(4z) - \frac{485919}{10240000}\Delta_{\text{[illegible]}}(6z) + \frac{17997}{2500}\Delta_{\text{[illegible]}}(12z) \\
& - \frac{97}{4992000000}\Delta_{\text{[illegible]}}(z) - \frac{97}{39000000}\Delta_{\text{[illegible]}}(2z) + \frac{297861}{1664000000}\Delta_{\text{[illegible]}}(3z) \\
& + \frac{297861}{13000000}\Delta_{\text{[illegible]}}(6z) - \frac{517}{23601217536}\Delta_{\text{[illegible]}}(z) - \frac{517}{327794688}\Delta_{\text{[illegible]}}(2z) \\
& - \frac{517}{720252}\Delta_{\text{[illegible]}}(4z) + \frac{1861}{96111360000}\Delta_{\text{[illegible]}}(z) + \frac{24193}{5339520000}\Delta_{\text{[illegible]}}(2z) \\
& + \frac{29776}{46929375}\Delta_{\text{[illegible]}}(4z) - \frac{11}{6523846656}\Delta_{\text{[illegible]}}(z) - \frac{598035}{2174615552}\Delta_{\text{[illegible]}}(3z) \\
& - \frac{77}{2723217408}\Delta_{\text{[illegible]}}(z) + \frac{77}{21275136}\Delta_{\text{[illegible]}}(2z) + \frac{1367}{59050819584}\Delta_{\text{[illegible]}}(z) \\
& - \frac{1367}{461334528}\Delta_{\text{[illegible]}}(2z) + \frac{1547}{110592000000}\Delta_{\text{[illegible]}}(z) \\
& + \frac{1547}{864000000}\Delta_{\text{[illegible]}}(2z) - \frac{3133}{50594512896}\Delta_{\text{[illegible]}}(z) \\
& + \frac{10111r + 98584027470}{3110720122965196800}\Delta_{\text{[illegible]}}(z) \\
& + \frac{-10111r + 99288359730}{3110720122965196800}\Delta_{\text{[illegible]}}(z).
\end{aligned}$$

$$\begin{aligned}
f_{\text{[illegible]}} = {} & \frac{47}{120960000}\Delta_{\text{[illegible]}}(2z) - \frac{1021}{15120000}\Delta_{\text{[illegible]}}(4z) + \frac{12771}{4480000}\Delta_{\text{[illegible]}}(6z) \\
& - \frac{36153}{560000}\Delta_{\text{[illegible]}}(12z) - \frac{23}{624000000}\Delta_{\text{[illegible]}}(2z) - \frac{354051}{208000000}\Delta_{\text{[illegible]}}(6z) \\
& - \frac{11}{61461504}\Delta_{\text{[illegible]}}(2z) + \frac{181}{7682688}\Delta_{\text{[illegible]}}(4z) - \frac{37}{250290000}\Delta_{\text{[illegible]}}(2z) \\
& - \frac{5461}{125145000}\Delta_{\text{[illegible]}}(4z) + \frac{1}{11031552}\Delta_{\text{[illegible]}}(2z) - \frac{5}{17036464}\Delta_{\text{[illegible]}}(2z) \\
& + \frac{17}{96000000}\Delta_{\text{[illegible]}}(2z)
\end{aligned}$$

$$
\begin{aligned}
f_{17} = &-\frac{87868564005559491556502096380642321549 9}{3682853172132649279085350129414446720000}\Delta_{1,18}(2z)-\frac{17148102875315060668515774406980627337693}{460356646516581159885668766176805840000}\Delta_{1,18}(4z)\\
&-\frac{76444138735308017471612629905993672326 6257}{136401969338246269595753708496831360000}\Delta_{1,18}(6z)+\frac{88557725372936042581343402818621621194 63101}{17050246167280783699469213562103920000}\Delta_{1,18}(12z)\\
&+\frac{53075774328418042315481049968024046596569}{5170989611563301573671819827437099312000000}\Delta_{2,18}(2z)-\frac{25198236125573394852496701289649817569 7053597}{56996537187767191223939942479033104000000}\Delta_{2,18}(6z)\\
&+\frac{2591875602769449724385620349959247929 99}{467827577237193105566099047867903717632}\Delta_{3,18,1}(2z)-\frac{176929274344290469699137335647704621408 7}{116956894309298276391524761966975929408}\Delta_{3,18,1}(4z)\\
&-\frac{71076473754876789323923478675866650281666 3}{9754299687334188233463198771334834598400 00}\Delta_{3,18,2}(2z)-\frac{72586214200056483796638296836978093625927939}{487714984366709411673159938566741729920000}\Delta_{3,18,2}(4z)\\
&+\frac{30108033285015100885446651141776646637}{27989684108208134521048660983549795072}\Delta_{4,18,1}(2z)-\frac{135958201607280717584912238482447969503}{346819544438548914967580400758571896832}\Delta_{4,18,2}(2z)\\
&-\frac{30559265484757601996764253053290796906477}{52612188173323561129790716134492096000000}\Delta_{4,18,3}(2z)+\frac{90291524865603541349728867831}{1325047702142963761787321739683504640}E_{18}(2z)\\
&-\frac{4482405405080632376282908663946101}{3975143106428891285361965219050513920}E_{18}(4z)+\frac{4357238443934985195238757084499917}{1325047702142963761787321739683504640}E_{18}(6z)\\
&-\frac{8589580801298920020057411776157143}{3975143106428891285361965219050513920}E_{18}(12z).
\end{aligned}
$$

$$
\begin{aligned}
f_{18} = &-\frac{177927375264736188626406926621830848 79}{982094179235373141089426701177185792 0000}\Delta_{1,18}(2z)-\frac{60426804669479809230256134348738970000 1}{409205908014738808787261125490494080000}\Delta_{1,18}(4z)\\
&-\frac{29167555593395337307004937806014706713147}{363738584901990052255343222658216960000}\Delta_{1,18}(6z)+\frac{82564496730812223539805579771600831471307}{151557743709162521773059676107590400000}\Delta_{1,18}(12z)\\
&+\frac{59355043019893654093443871391802795179 9}{455972297502137529791519539832264832000000}\Delta_{2,18}(2z)-\frac{780090182072919832861814107345983130012683 7}{151990765834045843263839846610754944000000}\Delta_{2,18}(6z)\\
&+\frac{20414146811668654807871193415833048 97}{2626400433612312171599152549433845432 32}\Delta_{3,18,1}(2z)-\frac{93942481148145134747179293332560827 89}{109433351400513007149964689559743559 68}\Delta_{3,18,1}(4z)\\
&-\frac{81446733498860047189527722330840994853 37}{6502866458222792155642132514223230656 0000}\Delta_{3,18,2}(2z)-\frac{30794699936547598136584149318876672749908 7}{1083811076370465359273688752370537177600 00}\Delta_{3,18,2}(4z)\\
&+\frac{81884548164102308273355946067593080 77}{42426468542968119695063233490854426214 4}\Delta_{4,18,1}(2z)-\frac{22045961439581124404486299502237056595}{62427517998938804694164472136542941429 76}\Delta_{4,18,2}(2z)\\
&-\frac{36852519067457706389839140532238694092 3}{35074792115549040753193810756328064000000}\Delta_{4,18,3}(2z)+\frac{90502396129158529305466933 9}{883365134761975841191547826455669760}E_{18}(2z)\\
&-\frac{292116171598447974862761350969 5}{176673026952395168238309565291133952}E_{18}(4z)-\frac{648371550387552097963989292448 11}{883365134761975841191547826455669760}E_{18}(6z)\\
&+\frac{794420585947163169542439421239 47}{883365134761975841191547826455669760}E_{18}(12z).
\end{aligned}
$$

$$
\begin{aligned}
f_{19} = &-\frac{171811732689426027511019076334251089537 3}{409205908014738808787261125490494080000}\Delta_{1,18}(2z)+\frac{18748429703159920733684137268006789090839}{51150738501842351098407640686311760000}\Delta_{1,18}(4z)\\
&-\frac{41980498791986460782602539997702647530888 9}{151557743709162521773059676107590400000}\Delta_{1,18}(6z)+\frac{57266865698580803087957440134620338531336 27}{18944717963645315221632459513448800000}\Delta_{1,18}(12z)\\
&-\frac{52187892132874314105286378229620058902387}{18998845729255730407979980826344368000000}\Delta_{2,18}(2z)+\frac{173269783499016256741428376995100913948108 81}{63329485764185768026599936087814560000 00}\Delta_{2,18}(6z)\\
&-\frac{14088540371069367554913323198859690054 7}{20792336766097471358493291016351276339 2}\Delta_{3,18,1}(2z)-\frac{43751675711831842354910848731606732104 83}{25990420957621839198116613770439095424}\Delta_{3,18,1}(4z)\\
&-\frac{66902063768040144707537539347361592601347}{13547638454630816990921109404631714720000}\Delta_{3,18,2}(2z)-\frac{59919821436260773567017750043996088406841 91}{67738192273154084954605547023158573600 00}\Delta_{3,18,2}(4z)\\
&+\frac{10965079688314094808036173601899351859 5}{11195873643283253808419464393419918028 8}\Delta_{4,18,1}(2z)-\frac{76078747466447625991206995777656558031}{21676221527409307185473775047410743552}\Delta_{4,18,2}(2z)\\
&-\frac{71395142622734376714498767915236569429 7}{18268120893515125392288443102254200000 0}\Delta_{4,18,3}(2z)+\frac{21602130308836605557255510 77}{920172015377058167907862319224656 0}E_{18}(2z)\\
&-\frac{2118265530648684340363652634584 53}{552103209226234900744717391534793 60}E_{18}(4z)-\frac{5873317909615535661610105283501}{230043003844264541976965579806164 0}E_{18}(6z)\\
&+\frac{7055464432349119759033468739120 3}{110420641845246980148943478306958 72}E_{18}(12z).
\end{aligned}
$$

REFERENCES

1. 1. Williams, K.S. (2012) Fourier Series of a Class of Eta Quotients. International Journal of Number Theory, 8, 993-1004.
2. Yao, O.X.M., Xia, E.X.W. and Jin, J. (2013) Explicit Formulas for the Fourier Coefficients of a Class of Eta Quotients. International Journal of Number Theory, 9, 487-503.
3. Köhler, G. (2011) Eta Products and Theta Series Identities. Springer-Verlag, Berlin.
4. Gordon, B. (1961) Some Identities in Combinatorial Analysis. Quarterly Journal of Mathematics, 12, 285-290.
5. Kac, V.G. (1978) Infinite-Dimensional Algebras, Dedekind's η-Function, Classical Möbius Function and the Very Strange Formula. Advances in Mathematics, 30, 85-136.
6. Macdonald, I.G. (1972) Affine Root Systems and Dedekind's η-Function. Inventiones Mathematicae, 15, 91-143.
7. Zucker, I.J. (1987) A Systematic Way of Converting Infinite Series into Infinite Products. Journal of Physics A, 20, L13-L17.
8. Zucker, I.J. (1990) Further Relations amongst Infinite Series and Products: II. The Evaluation of Three-Dimensional Lattice Sums. Journal of Physics A, 23, 117-132.
9. Kendirli, B. (2015) Evaluation of Some Convolution Sums by Quasimodular Forms. European Journal of Pure and Applied Mathematics, 8, 81-110.
10. Kendirli, B. (2015) Evaluation of Some Convolution Sums and Representation Numbers of Quadratic Forms of Discriminant -135. British Journal of Mathematics and Computer Science, 6, 494-531.
11. Kendirli, B. (2014) Evaluation of Some Convolution Sums and the Representation Numbers. Ars Combinatoria, CXVI, 65-91.

12. Kendirli, B. (2012) Cusp Forms in and the Number of Representations of Positive Integers by Some Direct Sum of Binary Quadratic Forms with Discriminant -79. Bulletin of the Korean Mathematical Society, 49, 529-572.

13. Kendirli, B. (2012) Cusp Forms in and the Number of Representations of Positive Integers by Some Direct Sum of Binary Quadratic Forms with Discriminant -47. International Journal of Mathematics and Mathematical Sciences, 2012, Article ID: 303492.

14. Kendirli, B. (2012) The Bases of, and the Number of Representations of Integers. Mathematical Problems in Engineering, 2013, Article ID: 695265.

15. Alaca, A., Alaca, S. and Williams, K.S. (2006) On the Two-Dimensional Theta Functions of Borweins. Acta Arithmetica, 124, 177-195.

16. Alaca, A., Alaca, S. and Williams, K.S. (2006) Evaluation of the convolution sums and . Advances in Theoretical and Applied Mathematics, 1, 27-48.

17. Gordon, B. and Robins, S. (1995) Lacunarity of Dedekind η-Products. Glasgow Mathematical Journal, 37, 1-14.

18. Diamond, F. and Shurman, J. (2005) A First Course in Modular Forms. Springer Graduate Texts in Mathematics 228. Springer, New York.

3

OPTIMAL SEPARATION OF TWIN CONVEX SETS UNDER EXTERNALITIES

Indrajit Mallick

Centre for Studies in Social Sciences, Kolkata, India

ABSTRACT

This paper studies the outcomes of independent and interdependent pair-wise contests between economic agents subject to an optimal external decision problem for each pair. The external decision maker like the government or regulator is faced with the problem of how to devise rules and regulations regarding contests. In this paper, a decision problem is faced under negative and positive externalities. A pair of entities is represented by disjoint convex sets in a small area in a neighborhood. I assume that each entity imposes an equal externality on the other (and the other only) and thus they can be considered to be twins. Among the group of twins in any neighborhood, there is a set of twin pairs such that, for each pair in the set, each twin can impose a strictly negative externality on the other (and the other only), and this is a potential welfare loss which concerns the decision maker. A separating hyper-plane can block the negative externalities between any pair of twins given convexity. However, this can be costly if positive externality from the neighborhood is also blocked by the separation technology. Thus, this paper compares the pair-wise utility from separation to that of non-separation. A simple representation of the decision problem is developed with respect to a single and isolated neighborhood. A complete characterization of the decision problem is obtained with a large number of pair-wise intersecting neighborhoods.

KEYWORDS

Convex Set, Separation, Externality

1. INTRODUCTION

Regulation and decentralization of institutional (legal rules and economic organization structure) setup are a primary challenge to improve the performance of an economy and bring about economic growth, financial stability and equitable wealth and income distribution. That an economy does not automatically achieve optimality with respect to these goals is by now well established (see in particular [1] - [4]). A regulator has to use some contingent intervention mechanisms in order to improve outcomes. All of these involve optimization under some basic assumptions about convexity.

Optimization and rules for decentralization in economic theory of choice and decision making rely heavily on convexity which naturally arises in the underlying preferences, budget sets, production sets and valuation sets. Maximization thus involves separating the set of benefits from the set of costs thus selecting the optimum while decentralization of resources uses the price system to separately allocate net benefits to economic units. However, additional complexity is introduced by a structure of externalities which leads to market failure unless corrected by some regulatory mechanisms. In this paper, I suggest such a mechanism under externalities by characterizing the decision problem of the regulator. It seems intuitively that when an economic entity makes a loss in engaging in games (of conflict, cooperation and coordination activities) with another entity due to negative externality imposed on each other, the loss could be minimized by not allowing the two entities to transact with each other. However, under positive externalities coming from some other source, separation decision might not be always optimal (except in case of outliers in a neighborhood). This is true both for independent and interdependent transactions and for both an isolated and a non-isolated neighborhood.

Consider the world of utility or value maximization by the individual units (like consumers, producers, traders, financial intermediaries, political groups (like political parties or political states) and social groups). Utility or value can be maximized through consumption, power, wealth and attainment of social objectives under economic, social and political interactions. Success is conditional on (besides own actions and resources) competing, cooperating and coordinating actions and resources of others. Thus selection of strategy by an agent is necessarily contingent on the strategies of others and the economic environment. Strategic contests thus characterize all economies, political systems and societies. A basic characteristic of strategic interaction is that utility (or value) maximizing economic agents are matched in a pair-wise fashion. A buyer is matched to a seller and a buyer may be matched to a number of sellers. With some loss of generality, it can be assumed that an economic unit like a buyer (who is a buyer in some matches and a seller in other matches) has multiple identities (with each identity relevant for each match) without some consolidating and general characteristics. The same can be assumed for a seller. Though pair-wise contests are potentially dependent on each other, independent pair-wise contests and the decision problems for regulators are first studied. This simplifies the analysis and the corresponding decision problem. I also discuss the implications of relaxing this restrictive assumption and derive a qualitatively similar result for the decision maker with interdependent pair-wise contests. The decision problem requires separation only when the sum of negative externalities dominates the sum of positive externalities.

Given that a buyer and a seller are matched, they engage in strategies of conflict like bargaining and contracting. Cooperation and coordination results if bargaining and contracting succeed and there is no cooperation and coordination if bargaining or contracting fails. Pair-wise contests are assumed to be driven by potential gains from trade and traditional rivalry with access to each other and where pairs are self-selected given their characteristics and their environments (see the analysis in [5] -[7]). This paper is based on a combination of economic phenomena which is generally observed in world where allocation of scarce resources is determined by pairwise decentralized competition given non-satiation in demand and the economy does not automatically attain the first best welfare given a structure of exogenously and endogenously determined externalities (see the discussion in [8] -[10]). Under these circumstances, it is well known that an appropriate system of regulation and an optimal design of a mix of centralization-decentralization of the institutional structure are required to achieve economic efficiency. This paper studies the optimal design of regulation and the required optimal mix of centralization and decentralization of the institutional and corresponding organization structure by an analysis of outcomes of pairwise contests between economic agents subject to an optimal external decision problem for each pair.

I first describe twin convex sets in a neighborhood. A pair of entities is represented by two disjoint convex sets (closed balls of same radius) in an arbitrarily small area of a neighborhood and a set of utility assignments associated with different types of externalities faced. The potential utility from positive externality to each of the twin convex sets is same. The centre of each set is equidistant to the centre of the neighborhood and each can potentially receive the same amount of positive externality from a ray coming from the centre. The potential externality can be interpreted as coming from public goods (see the exposition in [11] [12]), government programs (see the discussions in [13] -[19]) and coordination games of private activity with positive spillovers at the heart of the economy as the analysis in [4] [11] [14] [15] demonstrate.

I also assume that each entity can potentially impose an equal amount of externality on the other (and the other only). Thus they can be called twins. When a buyer and seller realize gains from trade they are essentially imposing non-negative externality on each other. This is satisfactory from the viewpoint of the decision maker and does not need any intervention. On the other hand, strictly negative externality imposed by a twin on the other and vice versa, is the essential welfare loss problem. It is assumed that the set of pairs of twins which realize gains from trade in a neighborhood $N(i)$ is denoted by the set $S(i)$ ($S(i)$

stands for success set) and the complement set in neighborhood $N(i)$ is denoted by $F(i)$ ($F(i)$ stands for the failure set).

There are pairs of twins which achieve gains from trade or gains from collusion or gains from coordination despite potential destructive competition or costs of coordination but these are not considered to be problems that affect a neighborhood. However, it is instructive to note how gains from trade are realized in pair-wise contests (see the analysis in [4] [10] [11] [15] [20] [21]). One or more of the following conditions are usually present: there is specific structure of specialization patterns, which through increasing productivity in activities creates a potential for gainful exchange, there is also the comparative advantage phenomena, differences in resource combinations and qualities, and the relative advantages of safe and risky assets. Each requires some degree of heterogeneity among entities. Entities can specialize in high risk but high coordination activities and insure against extreme specialization through risk sharing ex ante (see [15]). Positive externality is also realized in exchange when one economic entity creates a favorable situation for production or consumption of the counterparty and this increases the region of gains from trade. Of course in any of these situations, valuation to a buyer exceeds that of the seller.

The economics literature also discusses the reasons for negative externalities. One can start by assuming that the advantages in trade discussed above are missing and moreover different kinds of disadvantages manifest. But there is more than that. Preferences, endowments and production sets may be very different and generate negative externalities. The abstract analysis of standard negative externalities like pollution and encroachment are discussed in [8] [10] while [9] [10] discuss the costly contracting problems of moral hazard and adverse selection. The negative externalities of transaction costs like holdup and state verification problem which leads to suboptimal investment and under-provision of liquidity is analyzed in [22] . Incomplete information also results in bargaining inefficiency and is discussed in [7] [10] [15] . Other sources of negative externalities are absence of properly defined property rights (see the discussion in [8] [11] [22]), lack of incentives to invest and trade due to strategic uncertainty and unforeseen contingencies (see the analysis in [4] [11] [22] , imperfect information (discussed in [9] [10]) and envy (analyzed in[11]).

I shall focus on the problem of negative externality imposed by any twin on the corresponding twin in $F(i)$ subject to other considerations. I assume that each twin in $F(i)$ has a potential net loss of −n (n > 0) due to the negative externality that can be imposed by the other twin without separation. Given that for any twin-pair in $F(i)$, the twins are disjoint and convex, there exists a hyper-plane which can separate the twins. I assume that the hyper-plane is a technology which can completely block the negative externalities between any pair of twins. Thus it may seem that using the separation technology will be the right decision by the government regulating pair-wise contests. However, considerations become complicated because of the presence of the positive externality on each of the twins within its neighborhood coming from the centre of a neighborhood. Thus, this paper undertakes to analyze whether there is any need for separation technology when positive externality is blocked by the hyper-plane. A simple representation of the decision problem is developed with respect to a single neighborhood. A complete characterization of the decision problem is obtained with a large number of neighborhoods and overlapping positive externalities. This paper is more concerned about the desirability (normative analysis) of separation even if separation is feasible given the assumptions of convexity and disjoint twin sets (positive analysis). The focus of the paper is to first highlight the possible tradeoffs in the separation decision. The decision problem is characterized so that pair-wise utility value of separation is compared to pair-wise utility value of non-separation. The decision problem of the regulator is developed for independent as well as interdependent pair-wise contests where the latter is a generalization of the former with some restriction on separation decisions. At the same time, it should be noted that the present paper has a special approach which when generalized, could inform and improvise the decentralized interdependent bargaining model by introducing potential dynamics. The discussion about interdependent dynamic bargaining games is made in the conclusion where several suggestions for optimal regulation and decentralization decisions are developed for analyzing more complicated problems.

A classification of games is useful for a deeper understanding of economic phenomena and the need for regulation and redesign of economic systems. Economic entities compete for scarce resources that allow utility or value through consumption. As the analysis in [11] [15] [20] [21] suggest, at the basic level, potential competitors enter into costly conflicts with games of entry, predation, market share rivalry, takeover and attrition with exit resulting. In a pair-wise contest, conflict is always primary between two matched contestants. However, cooperation can emerge to a large extent due to self-preservation tendencies that equalizes the value of strategic options of being an aggressor and being timid. The credit for this insight goes to [23] who analyzes pair wise contests in an evolutionary biology context leading to the powerful concept of "Evolutionarily Stable Strategy".

The literature on dynamic matching and bargaining games is relevant for pair-wise matching contests. Different kinds of matchings (exogenous and endogenous) and assumptions about bargaining can be made. Both gain and loss is possible in these bargaining games and a general dynamic game should accommodate both. I start with a discussion of gains and cooperation. One significant notion of cooperation is potential gains from exchange through bargaining and contracting between a matched pair (though one should note that bargaining and contracting is potentially a game of conflict and cooperation). The result of such a game between a matched pair could have multiple solutions or exhibit indeterminacy as noted by [24] . However, as the number of agents grows sufficiently large and re-contracting possibilities are allowed, [24] shows the outcome to be the perfectly competitive equilibrium outcome with no indeterminacy. In the same spirit, dynamic matching and bargaining games are discussed by [6] [7] where games go over time and at any time, each of the simultaneous bargaining games is a pair-wise contest. Each bargaining depends on the future outcomes like exit if bargaining succeeds and re-matching and prospective value from future bargaining if bargaining fails at present. Bargained prices are shown by [6] to converge to the Walrasian price(s) even with dispersed characteristics provided discount factors approach unity. However, the decentralized pair-wise exchange process has several hazards which prevent its attainability and reduce its utility even if attainable. Costs of different exchange related phenomena like search, matching, learning to bargain, contract drafting and post-contractual unforeseen contingencies that are known as transaction costs are discussed by [25] [26] which are shown to impede trade. In the context of industrial organization, [21] shows that duopoly competition does not attain efficiency. Negative externalities discussed above lead to the pair dissolving and waiting for a favorable match under a dynamic game.

This paper is a brief attempt to characterize optimal regulation and institutional design. It anticipates a more general problem of regulation given the additional complexity introduced by interdependence and interdependence of pair-wise contests and given complexity induced by different possible kinds of dynamics. As such, a brief review of the literature on regulation is necessary. The problems of regulating a (potential) monopoly, inefficiently excessive competition, inefficient collusion and unstable coordination problems require the optimal design of regulation and are discussed in [5] [9] -[11] [15] [19] [21] . Standard devises suggested in the literature on industrial regulation are antitrust regulation, price control and control of entry. Other devices suggested by the literature are discriminating screening contracts devised by the government, vertical and horizontal integration and prudential regulation of banks. In the context of the problem of transaction costs associated with organizing economic activity, [8] has argued that, decentralized market solution is not necessarily optimal and the authority (which reduces transaction costs) based firm must play a relatively more important role. In an insightful analysis by [8] , it is argued that in particular, for the problem of externalities in the presence of transaction costs, the optimal solution should be the efficient allocation of property rights such that the efficient methods of countering the negative externality and allowing the positive externality are actually implemented. In discussing the optimal combination of centralization and decentralization in firms and economic systems, [27] has emphasized management and organization design as the principal coordination devices.

The historical development of relevant mathematical tools provides clues on the appropriate ways to approach the problem, the crucial assumptions needed, and the creative avenues that can be explored using optimization techniques, and the use of convexity. The first systematic study of the general properties of convexity and the separating hyper-plane theorem was due to [28] . A major work with comprehensive coverage on convex analysis is due to [29] which is still the basic text on the subject. The theoretical framework of linear economic models was developed by [30] who also provided analysis of two person games with other notable contemporary game theories and mathematicians and showed equivalence of two person games to appropriately redefined linear programming problems. Optimization by vector space methods was pioneered by [31] and allowed more productive use of linear algebra in optimization. The pioneering work by [32] elegantly analyzes the properties of convex sets and concave functions and provides an economic interpretation of separation through a hyperplane as optimization by decentralizing resources using a price system. A nice treatment of mathematical analysis especially in terms of basic topology, sequence and series convergence and properties of continuity has been provided in [33] .

2. THE DECISION PROBLEM FOR A SINGLE ISOLATED NEIGHBORHOOD

Given the assumptions about twin sets, the structure of externalities and the convexity assumptions, I discuss the decision problem both for independent and interdependent pair-wise contests. In particular, I discuss the relative merit and demerit of applying the hyper-plane theorem for the disjoint convex twin sets. First I describe the neighborhood and the externalities. I only discuss twins with negative externalities.

Assumption 2.1. Consider the Euclidean Space R^K. A neighborhood in R^K is an open set $N(i)$ where i is a positive integer and the centre of the set is $c(i)$ which has radius $r(i)$.

Assumption 2.2. I assume that each neighborhood $N(i)$ is isolated in the sense that there is no intersection between any of the neighborhoods.

Assumption 2.3. For set $S(i)$, any twin in a pair can receive a positive externality (equal to net gain in utility = m > 0) from the other twin in the pair, if non-separated. For set $F(i)$, any twin in a pair can receive a negative externality (equal to net loss in utility = −n < 0) from the other twin in the pair, if non-separated. The decision problem is for each twin-pair in $F(i)$.

Assumption 2.4. Let X and Y be two twin disjoint convex sets (closed balls of same radius belonging to set $F(i)$ in the neighborhood $N(i)$). The positive externality comes from the centre of the neighborhood and in order to have the positive utility effect on each of the twin pairs, it must reach the centre of each set representing the twins. The centre of each neighborhood emits a ray throughout the neighborhood with utility value equal to positive externality $A(i)-\mathrm{d}\left[c(i),c(X)\right]$ where $\mathrm{d}\left[c(i),c(X)\right]$ represents distance travelled by the ray from the centre of the neighborhood to the centre of any set X in the neighborhood. The maximum distance travelled by the ray is $r(i)$ such that $A(i)-r(i)=0$. In other words, the positive externality is exhausted when maximum distance equivalent to the radius is attained by a ray. If X and Y are twin sets then $\mathrm{d}\left[c(i),c(X)\right]=\mathrm{d}\left[c(i),c(Y)\right]$.

Assumption 2.5. Separation is a technology. Since any of a pair of twin sets are convex and disjoint, they can be separated using a separating hyper-plane (see [28] [29] [32]). I assume that a separating hyper-plane has real life separating properties in the sense that it blocks the negative externalities imposed by each set on the twin set and vice versa. This will ensure that as long as separation holds, there is no negative externality effect. I further assume that a separating hyper-plane is confined to the (arbitrarily) small area where any pair of the twin sets lie.

There are two ways of separation:

Case 1. There exists, and is optimally chosen, a separating hyper-plane line which meets the centre of the neighborhood from where the positive externality is coming. This ensures that negative externality can be removed without any effect on the positive externality since the hyper-plane does not block the path of positive externality on each set. In this case, separation is always an optimal decision.

Case 2. No separating hyper-plane line exists which meets the centre of the neighborhood. This is due to the fact that the twin sets are at an angle such that the separating line does not meet the centre of the neighborhood. This implies that due to separation, one set of the twins (let us call it Y) will not be a beneficiary of the positive externality coming from the centre of the neighborhood since the hyper-plane blocks the path of the positive externality. It does not matter which twin set will be denied positive externality, the total value of positive externality for the pair of sets is the same because the addition to utility from $A(i)-\mathrm{d}\left[c(i),c(X)\right]$ is constant over the two sets.

Assumption 2.6. Any ray from $c(i)$ potentially reaches only one convex twin set. This means that no other twin set (say Z or W) comes in the way of the positive externality on X and Y. Further, each set in the neighborhood potentially receives a ray from $c(i)$. Positive externality is generated from the centre of the neighborhood and is received by a set unless separation involves blocking of the positive externality coming to a twin set.

Assumption 2.7. In Case 2, if separation is considered to be optimal then $\left(A(i)-\mathrm{d}\left[c(i),c(X)\right]\right)/2$ will be allocated to each set.

2.1. Independent Pair-Wise Contests with Externalities

The decision criterion for Case 1. Separation is optimal.

The decision criterion for Case 2.

Value from separation:

$$V(s)=\left(A(i)-\mathrm{d}\left[c(i),c(X)\right]\right) \tag{1}$$

Value from non-separation:

$$V(ns)=\left(A(i)-\mathrm{d}\left[c(i),c(X)\right]-n\right)+\left(A(i)-\mathrm{d}\left[c(i),c(Y)\right]-n\right)=2\left(A(i)-\mathrm{d}\left[c(i),c(X)\right]-n\right) \tag{2}$$

Since the centre of each of the twin sets are equidistant from the centre of the neighborhood.

So non-separation is optimal iff: $\left(A(i)-\mathrm{d}\left[c(i),c(X)\right]\right)\geq 2n$.

2.2. Interdependent Pair-Wise Contests with Externalities

I consider only twin (disjoint and convex) sets with negative externalities in $F(i)$. Positive externalities as in $S(i)$ can be ignored without loss of generality. I therefore consider interdependent contests where each entity is represented in multiple contests with loss in each contest. Therefore the decision problem is with respect to the consolidated entity or the basic economic unit.

Assumption 2.2.1. Each twin set is denoted by $X(t(i))$ and the corresponding twin $Y(t(i))$ (where $t(i)=s(i)$ or $l(i)$ belonging to the set $S(i)$) or $F(i)$ respectively).

Assumption 2.2.2. For each $l(i)=1$ to L, an entity $X(l(i))$ is matched with another twin $Y(l(i))$ L times in set $F(i)$.

Assumption 2.2.3. For any consolidated entity or unit, separation means blocking negative externalities for all matches of the entities representing the unit in $F(i)$. In other words, discrimination in separation is not allowed since rules are made for the consolidated entity. However, separation decisions for negative externalities are not allowed to affect the gains from trade made by an economic unit in $S(i)$ (note that if separation implies separation for all matches both in $S(i)$ and $F(i)$, the case for non-separation is correspondingly stronger).

Assumption 2.2.4. In Case 2, if separation is considered to be optimal then $\left(A(i)-\mathrm{d}\left[c(i),c\left(X\left(l(i)\right)\right)\right]\right)/2$ will be allocated to each set.

The decision criterion for Case 1. Separation is optimal.

The decision criterion for Case 2.

Value from separation:

$$V(s)=\sum_{l(i)=1}^{L}\left(A(i)-\mathrm{d}\left[c(i),c\left(X\left(l(i)\right)\right)\right]\right) \tag{3}$$

Value from non-separation:

$$V(ns)=\sum_{l(i)=1}^{L}\left(A(i)-\mathrm{d}\left[c(i),c\left(X\left(l(i)\right)\right)\right]-n\right)+\sum_{l(i)=1}^{L}\left(A(i)-\mathrm{d}\left[c(i),c\left(Y\left(l(i)\right)\right)\right]-n\right)$$

or

$$V(ns)=\sum_{l(i)=1}^{L}\left(A(i)-\mathrm{d}\left[c(i),c\left(X\left(l(i)\right)\right)\right]-n\right)+\sum_{l(i)=1}^{L}\left(A(i)-\mathrm{d}\left[c(i),c\left(X\left(l(i)\right)\right)\right]-n\right) \tag{4}$$

Since the centre of each of the twin sets are equidistant from the centre of the neighborhood.

So non-separation is optimal iff: $\sum_{l(i)=1}^{L}\left(A(i)-\mathrm{d}\left[c(i),c\left(X\left(l(i)\right)\right)\right]\right)\geq 2Ln$.

3. THE DECISION PROBLEM FOR A LARGE NUMBER OF NON-ISOLATED NEIGHBORHOODS

The decision problem for a large number of neighborhoods is difficult and I avoid analysis of an unrestricted domain. Instead, I assume that neighborhoods have a hierarchical structure (in terms of increasing positive externalities $A(i)$ and radii $r(i)$ as i increases). This then allows a representation of the space covering the set of all neighborhoods as a cone C^K closed under scalar multiplication. The cone is convex given the following conditions hold: any of the (closed) set of points connected by a straight line between any two points in the cone belongs to a neighborhood and is nonempty (there are activities given externalities everywhere in any neighborhood) or belongs to the space in the cone which is not belonging to any neighborhood and is nonempty (there are isolated activities in each point in that space).

Assumption 3.1. The space that provides a complete cover for all the sets is a convex cone having the origin as the vertex. It is assumed that $c(i+1)$ is located to the northwest of set $c(i)$ for all i. Further, I assume that for all i, $A(i) < A(i+1)$, so that as one moves up northwest, positive externalities are stronger in the successive neighborhoods. Also, $r(i+1) = r(i) + \varepsilon$, where ε is an arbitrarily small positive number.

Assumption 3.2. $\mathrm{d}[c(i), c(i+1)] < [r(i) + r(i+1)]$ for all i, so intersection does occur between two adjoining neighborhoods. Further, for all i and $N(i)$, $N(i+1)$ and $N(i+2)$, there is a restriction on each radius such that the following condition holds: $r(i) < \mathrm{d}[c(i), c(i+1)]$ and $r(i+2) < \mathrm{d}[c(i+2), c(i+1)]$. This will ensure that for the analysis of the properties of intersection of two adjoining neighborhoods, no other neighborhood matters.

Assumption 3.3. For Case 2, if separation is considered to be optimal then $(A(i) - \mathrm{d}[c(i), c(X)])/2$ will be allocated to each set.

3.1. Independent Pair-Wise Contests with Externalities

The decision criterion for Case 1. Separation is optimal.

The decision criterion for Case 2. Let X and Y (a pair of disjoint convex twin sets) belong to the intersection of $N(i)$ and $N(i+1)$ for any i. Now, if the sets X and Y are separated, the positive externality from $c(i)$ hits one set and the positive externality from $c(i+1)$ hits the other set given the angularity assumption about separation.

Value from separation:

$$V(s)=\left(A(i)-\mathrm{d}\left[c(i),c(X)\right]\right)+\left(A(i+1)-\mathrm{d}\left(\left[c(i+1),c(Y)\right]\right)\right) \quad (5)$$

Value from non-separation:

$$V(ns)=\left(A(i)-\mathrm{d}\left[c(i),c(X)\right]\right)+\left(A(i+1)-\mathrm{d}\left[c(i+1),c(X)\right]\right)+\left(A(i+1)-\mathrm{d}\left[c(i+1),c(Y)\right]\right)$$
$$+\left(A(i)-\mathrm{d}\left[c(i),c(Y)\right]\right)-2n$$

So $V(ns)=2\left(A(i)-\mathrm{d}\left[\left(c(i),c(X)\right]\right)\right)+2\left(A(i+1)-\mathrm{d}\left[c(i+1),c(X)\right]\right)-2n$

(6)

Since the centre of each of the twin sets are equidistant from the centre of the relevant neighborhood.

So non-separation is optimal iff: $\left(A(i)-\mathrm{d}\left[c(i),c(X)\right]\right)+\left(A(i+1)-\mathrm{d}\left[c(i+1),c(X)\right]\right)\geq 2n$.

3.2. Interdependent Pair-Wise Contests with Externalities

The decision criterion for Case 1. Separation is optimal.

The decision criterion for Case 2.

Assumption 3.2.1. Let $X(l(i,i+1))$ be an economic unit in the intersection of $N(i)$ and $N(i+1)$ and let it be represented by $X(l(i))$ for $l(i)=1$ to $J<L$. The same applies for $Y(l(i,i+1))$ and $Y(l(i))$ where $X(l(i))$ and $Y(l(i))$ are twins for each $l(i)=1$ to $J<L$. Separation decision for any pair of twins imply separation decision between the two corresponding economic units also. Now, if the twin sets are separated, the positive externality from $c(i)$ hits one set and the positive externality from $c(i+1)$ hits the other set given the Case 2.

Assumption 3.2.2. In Case 2, if separation is considered to be optimal then $\left(A(i)-\mathrm{d}\left[c(i),c\left(X\left(l(i)\right)\right)\right]\right)/2$ will be allocated to each set.

Value from separation

$$V(s)=\sum_{l(i)=1}^{J}\left(A(i)-\mathrm{d}\left[c(i),c\left(X\left(l(i)\right)\right)\right]\right)+\sum_{l(i+1)=1}^{J}\left(A(i+1)-\mathrm{d}\left[c(i+1),c\left(Y\left(l(i+1)\right)\right)\right]\right) \tag{7}$$

Value from non-separation:

$$V(ns)=2\sum_{l(i)=1}^{J}\left(A(i)-\mathrm{d}\left[c(i),c\left(X\left(l(i)\right)\right)\right]\right)+2\sum_{l(i+1)=1}^{J}\left(A(i+1)-\mathrm{d}\left[c(i+1),c\left(X\left(l(i+1)\right)\right)\right]\right)-2Jn \tag{8}$$

So non-separation is optimal iff:

$$\sum_{l(i)=1}^{J}\left(A(i)-\mathrm{d}\left[c(i),c\left(X\left(l(i)\right)\right)\right]\right)+\sum_{l(i+1)=1}^{J}\left(A(i+1)-\mathrm{d}\left[c(i+1),c\left(X\left(l(i+1)\right)\right)\right]\right)\geq 2Jn$$

Since the centre of each of the twin sets are equidistant from the centre of the relevant neighborhood.

4. CONCLUSIONS

In this paper, decision criteria have been characterized with respect to the choice of separating or not separating disjoint twin convex sets in small areas of an isolated neighborhood where twins impose potential negative externality on each other and where positive externality comes from the centre of any neighborhood. Wherever there are tradeoffs between blocking negative externality using separation technology and receiving positive externality, the key issue is the distance of the ray travelled from the centre of the neighborhood. When the distance is low relative to other parameters, the positive externality is relatively strong and non-separation is considered to be optimal. On the other hand, when distance is high, positive externality is low and separation becomes optimal. Outliers (belonging far away from centre of a neighborhood) are better candidates for separation. Results are qualitatively same for independent and interdependent pair-wise contests.

I have also considered many neighborhoods in a convex cone under pair-wise intersection of neighborhoods assumption and the decision criteria have been similar but favoring non-separation relatively more since allowing positive externality from two neighborhoods is found more attractive. Otherwise, the distance travelled relative to other parameters is the crucial one. Results are qualitatively same for independent and interdependent pair-wise contests.

A generalization of the present analysis is possible by making the externalities endogenous and ensuring they change over time with each separation decision such that there are also inter-temporal tradeoffs. Another generalization is possible by considering the assumption of large number intersecting and interactive neighborhoods.

Optimization under convexity is an important tool for regulation as has been emphasized here. As mentioned in the beginning, in economic systems, three chief concerns in economic regulation are growth, stability and equitable distribution. Growth results from low capital-output ratios displayed by efficient technologies, technological progress and accumulation of human capital (see [4]). Instability of a growth path has been noted in some contexts. Financial instability has been discussed in [16] stressing asset risk due to moral hazard and excessive financial innovation (see[15] [25]) while the financial systemic financial fragility due to some salient properties of liquidity has been analyzed from several viewpoints in [1] [15] [25] . Inequality and poverty are analyzed and discussed in [4] [12] . Inequality is seen as stemming from unequal opportunities, results from lack of trust (see [4] [11]) through adverse selection and moral hazard, endogenously arise through missing markets and incomplete participation.

Despite important insights, there are problems in the foundations of this literature. Bilateral exchange is as important as organized exchange and even in the organized exchange there are ex ante pair-wise matchings between potential buyers and potential sellers under basic heterogeneities while there is ex post bilateral renegotiation given transaction costs, unforeseeable uncertainties and externalities. At the same time, bilateral bargaining and contracting games proceed in the context of varying and dispersed characteristics of buyers and sellers while histories of games of the past, current play and the future expectations link the bargaining games in an overall dynamic general equilibrium setup. Thus, sellers can be rationed in one class of decentralized pair-wise market leading to lack of effective demand in other decentralized pair-wise market. Other phenomena like delay in investments, slow accumulation of capital, transactions contingent on gradual learning and experimentation, evolution of trading strategies in financial markets and resulting financial innovations and low economic mobility might arise.

The analysis in this paper can be extended for the basic economic problems analyzed as dynamic pair-wise matching and bargaining games. For understanding growth, it is useful to start with the observation that intervention mechanisms for creation of surplus in bilateral bargaining games are necessary. This can be achieved through appropriate provision of public goods like network externalities mechanisms and technology but it should be recognized that fiscal and monetary constraints limit the possibilities. Interdependent pair-wise games can create positive spillovers given mechanisms of positive expectations generated by a policy maker (where the expectations are feasible and can be satisfied under a set of conditions requiring coordination and centralization) with respect to gains from trade. In this way, an increase in economic growth can be coordinated. Successful designs of complementary institutions for pair-wise games also result in endogenous provision of platforms of public goods (like the stock of knowledge, information and network technologies and financial innovations) thus reinforcing the growth process. Financial fragility and instability can be mitigated through certain indirect mechanisms. Regulation of banks, over-the-counter markets, financial innovations and corporate governance through combinations of dynamic screening contracts contingent rules and monitoring allow aggregative regulation as well as regulation of pair-wise matched individual entities which reduces financial fragility and instability. Regulation in the context of economic inequality and poverty can be designed in several ways. The provisions of need based liquidity and finance in combination with developing ability and incentives to repay create conditions for economic surplus and potential gains from trade, government programs for provision of public goods in health and education raises human capital and surplus in bilateral exchange, building trust in trading partners becomes possible in bilateral exchange by information sharing infrastructure and rules embedded in law, and the appropriate institution design of organizations and markets leads to efficient trading rules in bargaining games. Successful learning and experimentation in interdependent pair-wise contests given these interventions bring about the needed inequality reducing innovations.

ACKNOWLEDGEMENTS

I thank an anonymous reviewer for helpful comments and suggestions.

REFERENCES

1. Allen, F. and Douglas, G. (2007) Understanding Financial Crises. Oxford University Press, Oxford.

2. Hahn, F. (1985) Money, Growth and Stability. The MIT Press, Cambridge.

3. Keynes, J.M. (1936) The General Theory of Employment, Interest and Money. Macmillan, London.

4. Ray, D. (1999) Development Economics. Oxford University Press, New Delhi.

5. Fudenberg, D. and Jean, T. (2005) Game Theory. Ane Books, India.

6. Gale, D. (2000) Strategic Foundations of General Equilibrium: Dynamic Matching and Bargaining Games. Cambridge University Press, Cambridge.

7. Osborne, M.J. and Ariel, R. (1990) Bargaining and Markets. Academic Press, Inc., San Diego.

8. Coase, R.H. (1988) The Firm, the Market and the Law. The University of Chicago Press, Chicago.

9. Laffont, J.-J. (1989) The Economics of Uncertainty and Information. The MIT Press, Cambridge.

10. Mas-Colell, A., Whinston, M.D. and Green, J.R. (1995) Microeconomic Theory. Oxford University Press, Oxford.

11. Bowles, S. (2005) Microeconomics: Behavior, Institutions and Evolution. Oxford University Press, New Delhi.

12. Hayami, Y. (2001) Development Economics: From the Poverty to the Wealth of Nations. Oxford University Press, Oxford

13. Azariadis, C. (1993) Intertemporal Macroeconomics. Blackwell Publishers, Cambridge.

14. Blanchard, O.J. and Fischer, S. (1989) Lectures on Macroeconomics. The MIT Press, Cambridge.

15. Mallick, I. (2002) Strategic Competition in Banking: Theory and Policy. Ph.D. Thesis, Jadavpur University, Kolkata.

16. Dewatripont, M. and Tirole, J. (1993) The Prudential Regulation of Banks. The MIT Press, Cambridge.

17. Gale, D. (1983) Money: In Disequilibrium. Cambridge University Press, Cambridge.

18. Stiglitz, J.E. (1986) Economics of the Public Sector. W. W. Norton & Company, New York.

19. Viscusi, W.K., Harrington, J.E. and Vernon, J.M. (2005) Economics of Regulation and Antitrust. The MIT Press, Cambridge.

20. Bernard, M. (2012) Conflict, Cooperation and Coordination: Essays in Game Theory and Experimental Economics. Ph.D. Thesis, Stockholm School of Economics, Stockholm.

21. Tirole, J. (1988) The Theory of Industrial Organization. The MIT Press, Cambridge.

22. Hart, O. (1995) Firms, Contracts, and Financial Structure. Oxford University Press, Oxford.

23. Smith, J.M. (1982) Evolution and the Theory of Games. Cambridge University Press, Cambridge.

24. Edgeworth, F.Y. (1881) Mathematical Psychics: An Essay on the Application of Mathematics to the Moral Sciences. Kegan Paul, London.

25. Duffie, D. (2012) Dark Markets: Asset Pricing and Information Transmission in Over-the-Counter Markets. 3rd Edition, Princeton University Press, Princeton.

26. Shavell, S. (2004) Foundations of Economic Analysis of Law. Harvard University Press, Cambridge.

27. Milgrom, P. and Roberts, J. (1992) Economics, Organization and Management. Prentice-Hall, Inc., Upper Saddle River.

28. Minkowski, H. (1911) Theorie der Konvexen Korper, Insbesondere Begrundung ihres Oberflachenbegriffs. Gessamelte Abhandlungen II, Leipzig.

29. Rockafellar, R.T. (1970) Convex Analysis. Princeton University Press, Princeton.

30. Gale, D. (1960) The Theory of Linear Economic Models. McGraw-Hill, Inc., New York.

31. Luenberger, D.G. (1969) Optimization by Vector Space Methods. Wiley, New York.

32. Dixit, A.K. (1990) Optimization in Economic Theory. Oxford University Press, Oxford.

33. Rudin, W. (1976) Principles of Mathematical Analysis. McGraw-Hill, Inc., London.

4

A PARAMETER-DEPENDENT REFINEMENT OF THE DISCRETE JENSEN'S INEQUALITY FOR CONVEX AND MID-CONVEX FUNCTIONS

László Horváth

ABSTRACT

In this paper, a new parameter-dependent refinement of the discrete Jensen's inequality is given for convex and mid-convex functions. The convergence of the introduced sequences is also studied. One of the proofs requires an interesting convergence theorem with probability theoretical background. We apply the results to define some new quasi-arithmetic and mixed symmetric means and study their monotonicity and convergence.

1 INTRODUCTION AND THE MAIN RESULTS

The considerations of this paper concern

(A_1) an arbitrarily given real vector space X, a convex subset C of X, and a finite subset $\{x_1,..., x_n\}$ of C, where $n \geq 1$ is fixed;

(A_2) a convex function $f: C \to \mathbb{R}$, and a discrete distribution $p_1,..., p_n$, which means that $p_j \geq 0$ with $\sum_{j=1}^{n} p_j = 1$;

(A_3) a mid-convex function $f: C \to \mathbb{R}$, and a discrete distribution $p_1,..., p_n$ with rational p_j $(1 \leq j \leq n)$.

The function $f: C \to \mathbb{R}$ is called convex if

$$f(\alpha x + (1 - \alpha)y) \leq \alpha f(x) + (1 - \alpha)f(y), \quad x, y \in C, \quad 0 \leq \alpha \leq 1, \tag{1}$$

and mid-convex if

$$f\left(\frac{x+y}{2}\right) \leq \frac{1}{2}f(x) + \frac{1}{2}f(y), \quad x, y \in C.$$

For a variety of applications, the discrete Jensen's inequalities are important:

Theorem A. (see [1]) (a) *If (A_1) and (A_2) are satisfied, then*

$$f\left(\sum_{j=1}^{n} p_j x_j\right) \leq \sum_{j=1}^{n} p_j f(x_j). \tag{2}$$

(b) *If (A_1) and (A_3) are satisfied, then (2) also holds.*

Let $\mathbb{N} := \{0, 1, 2,...\}$ and let $\mathbb{N}_+ := \{1, 2,...\}$.

Various attempts have been made to refine inequality (2) in the following ways: Assume either (A_1) and (A_2) or (A_1) and (A_3). Let $m \geq 2$ be an integer, and let I denote either the set $\{1,..., m\}$ or the set $\mathbb{N}_+$.

(B) Create a decreasing real sequence $(B_k)_{k \in I}$ such that $B_k = B_k(f, x_i, p_i)$ $(k \in I)$ is a sum whose index set is a subset of $\{1,..., n\}^k$ and

$$f\left(\sum_{j=1}^{n} p_j x_j\right) \leq \cdots \leq B_k \leq \cdots \leq B_1 = \sum_{j=1}^{n} p_j f(x_j), \quad k \in I. \tag{3}$$

(C) Create an increasing real sequence $(C_k)_{k \in I}$ such that $C_k = C_k(f, x_i, p_i)$ $(k \in I)$ is a sum whose index set is a subset of $\{1,...,k\}^n$ and

$$f\left(\sum_{j=1}^{n} p_j x_j\right) = C_1 \leq \cdots \leq C_k \leq \cdots \leq \sum_{j=1}^{n} p_j f(x_j), \quad k \in I. \tag{4}$$

The next two typical results belong to the group of refinements of type (B).

These examples use $p_1 = \cdots = p_n = \frac{1}{n}$. In [2], Pečarić and Volenec have constructed the sequence

$$f_k := \frac{1}{\binom{n}{k}} \sum_{1 \leq i_1 < \cdots < i_k \leq n} f\left(\frac{x_{i_1} + \cdots + x_{i_k}}{k}\right), \quad 1 \leq k \leq n, \tag{5}$$

while the other sequence

$$\bar{f}_k := \frac{1}{\binom{n+k-1}{k}} \sum_{1 \leq i_1 \leq \cdots \leq i_k \leq n} f\left(\frac{x_{i_1} + \cdots + x_{i_k}}{k}\right), \quad k \in \mathbb{N}_+ \tag{6}$$

is due to Pečarić and Svrtan [3]. In a recent work, [4] Horváth and Pečarić define a lot of new sequences, they generalize and give a uniform treatment a number of well-known results from this area, especially (5) and (6) are extended. Horváth develops a method in [5] to construct decreasing real sequences satisfying (3). His paper contains some improvements of the results in [4] and gives a new approach of the topic. The description of the sequences in [4, 5] requires some work, so we do not go into the details. The problem (B) has been considered for the classical Jensen's inequality by Horváth [6].

We turn now to the group of refinements of type (C). In contrast to the previous problem, it is not easy to find such results. Recently, Xiao et al. [7] have obtained the sequence

$$F_k := \frac{1}{\binom{n+k-2}{k-1}} \sum_{\substack{i_1 + \cdots + i_n = n+k-1 \\ i_j \in \mathbb{N}_+ \, (1 \leq j \leq n)}} f\left(\frac{1}{n+k-1} \sum_{j=1}^{n} i_j x_j\right), \quad k \in \mathbb{N}_+, \tag{7}$$

which satisfies (4) with $p_1 = \cdots = p_n = \frac{1}{n}$.

In this paper, we establish a new solution of the problem (C). The constructed sequence $(C_k(\lambda))_{k\geq 0}$depends on a parameter λbelonging to $[1, \infty[$, and we can use arbitrary discrete distribution $p_1,..., p_n$, not just the appropriate discrete uniform distribution. We give the limit of the sequence under fixed parameter. We also study the convergence of the sequence when the parameter varies and $k \in \mathbb{N}$ is fixed. Finally, some applications are given which concern the theme of means.

The next theorems are the main results of this paper. We need some further hypotheses:

(A_4) Let $\lambda \geq 1$.

(A_5) Let $\lambda \geq 1$ be rational.

First, we give a refinement of the discrete Jensen's inequality (2).

Theorem 1 *Suppose either (A_1), (A_2), and (A_4) or (A_1), (A_3), and (A_5). Introduce the sets*

$$S_k := \left\{ (i_1, \ldots, i_n) \in \mathbb{N}^n \middle| \sum_{j=1}^{n} i_j = k \right\}, \quad k \in \mathbb{N},$$

and for $k \in \mathbb{N}$ define the numbers

$$C_k(\lambda) = C_k(x_1, \ldots, x_n; p_1, \ldots, p_n; \lambda)$$

$$:= \frac{1}{(n+\lambda-1)^k} \sum_{(i_1,\ldots,i_n)\in S_k} \frac{k!}{i_1! \ldots i_n!} \left(\sum_{j=1}^{n} \lambda^{i_j} p_j \right) f \left(\frac{\sum_{j=1}^{n} \lambda^{i_j} p_j x_j}{\sum_{j=1}^{n} \lambda^{i_j} p_j} \right). \quad (8)$$

Then,

$$f\left(\sum_{j=1}^{n} p_j x_j \right) = C_0(\lambda) \leq C_1(\lambda) \leq \cdots \leq C_k(\lambda) \leq \cdots \leq \sum_{j=1}^{n} p_j f(x_j), \quad k \in \mathbb{N}.$$

Remark 2 *(a) It follows from the definition of S_k that $S_k \subset \{0,..., k\}^n$ $(k \in \mathbb{N})$.*

(b) It is easy to see that

$$C_k(1) = f\left(\sum_{j=1}^{n} p_j x_j\right), \quad k \in \mathbb{N}. \tag{9}$$

Finally, we establish two convergence theorems.

Theorem 3 *Suppose (A_1), (A_2), and (A_4). Suppose × is a normed space and f is continuous. Then,*

(a) For every fixed λ > 1

$$\lim_{k\to\infty} C_k(\lambda) = \sum_{j=1}^{n} p_j f(x_j).$$

(b) The function $\lambda \to C_k(\lambda)$ ($\lambda \geq 1$) is continuous for every $k \in \mathbb{N}$.

The proof of Theorem 3(a) requires a lemma (see Lemma 15), which is interesting in its own right. Probability theoretical technique will be used to handle this problem.

Remark 4 *In the previous theorem, it suffices to consider the case when (A_1), (A_2), and (A_4) are satisfied. Really, if f is mid-convex and continuous, then convex.*

By (9)

$$\lim_{k\to\infty} C_k(1) = f\left(\sum_{j=1}^{n} p_j x_j\right).$$

We come now to the second convergence theorem.

Theorem 5 *Suppose either (A_1), (A_2), and (A_4) or (A_1), (A_3), and (A_5). For each fixed $k \in \mathbb{N}_+$*

$$\lim_{\lambda\to\infty} C_k(\lambda) = \sum_{j=1}^{n} p_j f(x_j).$$

2 DISCUSSION AND APPLICATIONS

Suppose either (A_1), (A_2), and (A_4) or (A_1), (A_3), and (A_5). First, we give three special cases of (8).

(a) $k = 1, n \in \mathbb{N}_+$:

$$C_1(\lambda) = \frac{1}{n+\lambda-1}\sum_{i=1}^{n}(1+(\lambda-1)p_i)f\left(\frac{\sum_{j=1}^{n} p_j x_j + (\lambda-1)p_i x_i}{1+(\lambda-1)p_i}\right).$$

(b) $k \in \mathbb{N}, n = 2$:

$$C_k(\lambda) = \frac{1}{(\lambda+1)^k}\sum_{i=0}^{k}\binom{k}{i}\left(\lambda^i p_1 + \lambda^{k-i}p_2\right) f\left(\frac{\lambda^i p_1 x_1 + \lambda^{k-i}p_2 x_2}{\lambda^i p_1 + \lambda^{k-i}p_2}\right).$$

(c)

$$p_1 = \cdots = p_n := \frac{1}{n}$$

$$C_k(\lambda) = \frac{1}{n(n+\lambda-1)^k}\sum_{(i_1,\ldots,i_n)\in S_k}\frac{k!}{i_1!\ldots i_n!}\left(\sum_{j=1}^{n}\lambda^{i_j}\right) f\left(\frac{\sum_{j=1}^{n}\lambda^{i_j}x_j}{\sum_{j=1}^{n}\lambda^{i_j}}\right).$$

Assume further that f is strictly convex (strictly mid-convex) that is strict inequality holds in (1) whenever $x \neq y$ and $0 < \alpha < 1$. In this case, equality is satisfied in (2) if and only if $x_1 = \cdots = x_n$, and therefore, it comes from the third part of the proof of Theorem 1 that

$$C_k(\lambda) < \sum_{j=1}^{n} p_j f(x_j), \quad k \in \mathbb{N}, \tag{10}$$

if not all x_i are equal.

If $p_1 = \cdots = p_n := \frac{1}{n}$ and f is strictly convex (strictly mid-convex), then the analysis of the proof of Theorem 1 shows that

$$f\left(\frac{1}{n}\sum_{j=1}^{n} x_j\right) = C_0(\lambda) < C_1(\lambda) < \cdots < C_k(\lambda) < \cdots < \frac{1}{n}\sum_{j=1}^{n} f(x_j), \quad k \in \mathbb{N},$$

whenever not all x_i are equal.

If the inequality (10) holds, X is a normed space and f is continuous (see Remark 4), then Theorem 3(b) and Theorem 5 insure that the range of the function $\lambda \to C_k(\lambda)$ ($k \in \mathbb{N}_+$) is the interval

$$\left[f\left(\sum_{j=1}^{n} p_j x_j\right), \quad \sum_{j=1}^{n} p_j f(x_j)\right[.$$

Conjecture 6 *Suppose either (A_1), (A_2), and (A_4) or (A_1), (A_3), and (A_5).*

The function $\lambda \to C_k(\lambda)$ $(\lambda \geq 1)$ is increasing for every $k \in \mathbb{N}$.

Next, we define some new quasi-arithmetic means and study their monotonicity and convergence. About means see [8].

Definition 7 *Let $I \subset \mathbb{R}$ be an interval, let $x_j \in I$ $(1 \leq j \leq n)$, let $p_1,\ldots, p_n$ be a discrete distribution, and let g, $h : I \to \mathbb{R}$ be continuous and strictly monotone functions. Let $\lambda \geq 1$. We define the quasi-arithmetic means with respect to (8) by*

$$M_{h,g}(k, \lambda) := h^{-1}\left(\frac{1}{(n+\lambda-1)^k} \sum_{(i_1,\ldots,i_n)\in S_k} \frac{k!}{i_1!\ldots i_n!}\left(\sum_{j=1}^{n} \lambda^{i_j} p_j\right) \cdot (h \circ g^{-1})\left(\frac{\sum_{j=1}^{n} \lambda^{i_j} p_j g(x_j)}{\sum_{j=1}^{n} \lambda^{i_j} p_j}\right)\right), \quad k \in \mathbb{N}. \tag{11}$$

Some other means are also needed.

Definition 8 *Let $I \subset \mathbb{R}$ be an interval, let $x_j \in I$ $(1 \leq j \leq n)$, and let $p_1,\ldots, p_n$ be a discrete distribution. For a continuous and strictly monotone function $z : I \to \mathbb{R}$, we introduce the following mean*

$$M_z := z^{-1}\left(\sum_{j=1}^{n} p_j z(x_j)\right). \tag{12}$$

We now prove the monotonicity of the means (11) and give limit formulas.

Proposition 9 *Let $I \subset \mathbb{R}$ be an interval, let $x_j \in I$ $(1 \leq j \leq n)$, let $p_1, \ldots, p_n$ be a discrete distribution, and let g, $h : I \to \mathbb{R}$ be continuous and strictly monotone functions. Let $\lambda \geq 1$. Then,*

(a) $M_g = M_{h,g}(0, \lambda) \leq \cdots \leq M_{h,g}(k, \lambda) \leq \cdots \leq M_h, \quad k \in \mathbb{N},$

if either $h \circ g^{-1}$is convex and h is increasing or $h \circ g^{-1}$is concave and h is decreasing.

(b) $M_g = M_{h,g}(0, \lambda) \geq \cdots \geq M_{h,g}(k, \lambda) \geq \cdots \geq M_h, \quad k \in \mathbb{N},$

if either $h \circ g^{-1}$is convex and h is decreasing or $h \circ g^{-1}$is concave and h is increasing.

(c) Moreover, in both cases

$$\lim_{k \to \infty} M_{h,g}(k, \lambda) = M_h$$

for each fixed $\lambda > 1$, and

$$\lim_{\lambda \to \infty} M_{h,g}(k, \lambda) = M_h$$

for each fixed $k \in \mathbb{N}_+$.

Proof. Theorem 1 can be applied to the function $h \circ g^{-1}$, if it is convex (-$h \circ g^1$, if it is concave) and the n-tuples $(g(x_1), \ldots, g(x_n))$, then upon taking h^{-1}, we get (a) and (b). (c) comes from Theorems 3(a) and 5. ■

As a special case, we consider the following example.

Example 10 *If* $I :=]0, \infty[$, $h :=$ ln, *and* $g(x) := x$ $(x \in]0, \infty[)$, *then by Proposition 9(b), we have the following inequality.for every $x_j > 0$ $(1 \leq j \leq n)$, $\lambda \geq 1$, and $k \in \mathbb{N}_+$*

$$\sum_{j=1}^{n} p_j x_j \geq \prod_{(i_1,\ldots,i_n)\in S_k} \left(\frac{\sum_{j=1}^{n} \lambda^{i_j} p_j x_j}{\sum_{j=1}^{n} \lambda^{i_j} p_j} \right)^{\frac{1}{(n+\lambda-1)^k} \frac{k!}{i_1!\ldots i_n!} \sum_{j=1}^{n} \lambda^{i_j} p_j} \geq \prod_{j=1}^{n} x_j^{p_j},$$

which gives a sharpened version of the arithmetic mean - geometric mean inequality

$$\frac{1}{n}\sum_{j=1}^{n} x_j \geq \prod_{(i_1,\ldots,i_n)\in S_k} \left(\frac{\sum_{j=1}^{n} \lambda^{i_j} x_j}{\sum_{j=1}^{n} \lambda^{i_j}} \right)^{\frac{1}{n(n+\lambda-1)^k} \frac{k!}{i_1!\ldots i_n!} \sum_{j=1}^{n} \lambda^{i_j}} \geq \prod_{j=1}^{n} x_j^{\frac{1}{n}}.$$

Finally, we investigate some mixed symmetric means.

The power means of order $r \in \mathbb{R}$ are defined as follows:

Definition 11 *Let* $x_j \in]0, \infty[$ $(1 \leq j \leq n)$, *and let* $p_1,\ldots, p_n$ *be a discrete distribution with* $p_j > 0$ $(1 \leq j \leq n)$.

$$M_r = M_r(x_1, \ldots, x_n; p_1, \ldots, p_n) := \begin{cases} \left(\sum_{j=1}^{n} p_j x_j^r\right)^{\frac{1}{r}}, & r \neq 0 \\ \left(\prod_{j=1}^{n} x_j^{p_j}\right), & r \neq 0 \end{cases}.$$

If $r \neq 0$, then the power means of order r belong to the means (12) ($z :]0, \infty[\to \mathbb{R}$, $z(x) := x^r$), while we get the power means of order 0 by taking limit. Supported by the power means, we can introduce mixed symmetric means corresponding to (8):

Definition 12 *Let* $x_j \in]0, \infty[$ $(1 \leq j \leq n)$, *and let* $p_1,\ldots, p_n$ *be a discrete distribution with* $p_j > 0$ $(1 \leq j \leq n)$. *Let* $\lambda \geq 1$, *and* $k \in \mathbb{N}$. *We define the mixed symmetric means with respect to (8) by*

$$M_{s,t}(k,\lambda)$$
$$:= \left(\frac{1}{(n+\lambda-1)^k} \sum_{(i_1,\dots,i_n)\in S_k} \frac{k!}{i_1!\dots i_n!} \left(\sum_{j=1}^{n} \lambda^{i_j} p_j \right) \cdot M_t^s \left(x_1,\dots,x_n; \frac{\lambda^{i_1} p_1}{\sum_{j=1}^{n} \lambda^{i_j} p_j}, \dots, \frac{\lambda^{i_n} p_n}{\sum_{j=1}^{n} \lambda^{i_j} p_j} \right) \right)^{\frac{1}{s}},$$

if $s, t \in \mathbb{R}$ *and* $s \neq 0$, *and*

$$M_{0,t}(k,\lambda) := \prod_{(i_1,\dots,i_n)\in S_k} \left(M_t \left(x_1,\dots,x_n; \frac{\lambda^{i_1} p_1}{\sum_{j=1}^{n} \lambda^{i_j} p_j}, \dots, \frac{\lambda^{i_n} p_n}{\sum_{j=1}^{n} \lambda^{i_j} p_j} \right) \right)^{\frac{1}{(n+\lambda-1)^k} \frac{k!}{i_1!\dots i_n!} \left(\sum_{j=1}^{n} \lambda^{i_j} p_j \right)},$$

where $t \in \mathbb{R}$.

The monotonicity and the convergence of the previous means are studied in the next result.

Proposition 13 *Let* $x_j \in]0,\infty[$ $(1 \leq j \leq n)$, *and let* $p_1,\dots,p_n$ *be a discrete distribution with* $p_j > 0$ $(1 \leq j \leq n)$. *Let* $\lambda \geq 1$, *and* $k \in \mathbb{N}$. *Suppose* $s, t \in \mathbb{R}$ *such that* $s \leq t$. *Then,*

(a) $M_t = M_{s,t}(0,\lambda) \geq \dots \geq M_{s,t}(k,\lambda) \geq \dots \geq M_s, \quad k \in \mathbb{N}.$

(b) In case of $s, t \neq 0$

$$\lim_{k\to\infty} M_{s,t}(k,\lambda) = M_s$$

for each fixed $\lambda > 1$, *and*

$$\lim_{\lambda\to\infty} M_{s,t}(k,\lambda) = M_s$$

for each fixed $k \in \mathbb{N}_+$.

Proof. Assume $s, t \neq 0$. Then, Proposition 9 (b) can be applied with $g, h :]0, \infty[\to \mathbb{R}$, $g(x) := x^t$, and $h(x) := x^s$. If $s = 0$ or $t = 0$, the result follows by taking limit.

3 SOME LEMMAS AND THE PROOFS OF THE MAIN RESULTS

Lemma 14 *Let $k \in \mathbb{N}$ and $(i_1,..., i_n) \in S_{k+1}$ be fixed. If we set*

$$z(i_1, \ldots, i_n) := \{j \in \{1, \ldots, n\} | i_j \neq 0\},$$

then

$$\sum_{j \in z(i_1,\ldots,i_n)} \frac{k!}{i_1! \ldots i_{j-1}!(i_j - 1)! i_{j+1}! \ldots i_n!} = \frac{(k+1)!}{i_1! \ldots i_n!}.$$

Proof. The lowest common denominator is $i_1! \ldots i_n!$. Combined with $\sum_{j=1}^{n} i_j = k + 1$, the result follows. ■

The proof of Theorem 1.

Proof. (a) We separate the proof of this part of the theorem into three steps.

Let $\lambda \geq 1$ be fixed.

I. Since $S_0 = \{(0,..., 0)\}$

$$C_0(\lambda) = \left(\sum_{j=1}^{n} \lambda^0 p_j\right) f\left(\frac{\sum_{j=1}^{n} \lambda^0 p_j x_j}{\sum_{j=1}^{n} \lambda^0 p_j}\right) = f\left(\sum_{j=1}^{n} p_j x_j\right).$$

II. Next, we prove that $C_k(\lambda) \leq C_{k+1}(\lambda)$ $(k \in \mathbb{N})$.

It is easy to check that for every $(i_1,..., i_n) \in S_k$

$$\frac{\sum_{j=1}^{n} \lambda^{i_j} p_j x_j}{\sum_{j=1}^{n} \lambda^{i_j} p_j} = \frac{1}{n+\lambda-1}$$

$$\cdot \sum_{l=1}^{n} \left(\frac{\sum_{j=1}^{n} \lambda^{i_j} p_j x_j + (\lambda-1)\lambda^{i_l} p_l x_l}{\sum_{j=1}^{n} \lambda^{i_j} p_j + (\lambda-1)\lambda^{i_l} p_l} \cdot \frac{\sum_{j=1}^{n} \lambda^{i_j} p_j + (\lambda-1)\lambda^{i_l} p_l}{\sum_{j=1}^{n} \lambda^{i_j} p_j} \right).$$

With the help of Theorem A, this yields that

$$f\left(\frac{\sum_{j=1}^{n} \lambda^{i_j} p_j x_j}{\sum_{j=1}^{n} \lambda^{i_j} p_j} \right) \leq \frac{1}{n+\lambda-1} \sum_{l=1}^{n} \left(\frac{\sum_{j=1}^{n} \lambda^{i_j} p_j + (\lambda-1)\lambda^{i_l} p_l}{\sum_{j=1}^{n} \lambda^{i_j} p_j} \right.$$

$$\left. \cdot f\left(\frac{\sum_{j=1}^{n} \lambda^{i_j} p_j x_j + (\lambda-1)\lambda^{i_l} p_l x_l}{\sum_{j=1}^{n} \lambda^{i_j} p_j + (\lambda-1)\lambda^{i_l} p_l} \right) \right).$$

Consequently,

$$C_k(\lambda) \leq \frac{1}{(n+\lambda-1)^{k+1}} \sum_{(i_1,\ldots,i_n)\in S_k} \frac{k!}{i_1!\ldots i_n!}$$

$$\cdot \sum_{l=1}^{n} \left(\left(\sum_{j=1}^{n} \lambda^{i_j} p_j + (\lambda-1)\lambda^{i_l} p_l \right) f\left(\frac{\sum_{j=1}^{n} \lambda^{i_j} p_j x_j + (\lambda-1)\lambda^{i_l} p_l x_l}{\sum_{j=1}^{n} \lambda^{i_j} p_j + (\lambda-1)\lambda^{i_l} p_l} \right) \right). \tag{13}$$

By Lemma 14, it is easy to see that the right-hand side of (13) can be written in the form

$$\frac{1}{(n+\lambda-1)^{k+1}} \sum_{(i_1,\ldots,i_n)\in S_{k+1}} \frac{(k+1)!}{i_1!\ldots i_n!} \left(\sum_{j=1}^{n} \lambda^{i_j} p_j \right) f\left(\frac{\sum_{j=1}^{n} \lambda^{i_j} p_j x_j}{\sum_{j=1}^{n} \lambda^{i_j} p_j} \right)$$

which is just $C_{k+1}(\lambda)$.

III. Finally, we prove that

$$C_k(\lambda) \le \sum_{j=1}^{n} p_j f(x_j), \quad k \in \mathbb{N}_+. \tag{14}$$

It follows from Theorem A that

$$C_k(\lambda) \le \frac{1}{(n+\lambda-1)^k} \sum_{(i_1,\dots,i_n)\in S_k} \left(\frac{k!}{i_1!\dots i_n!} \sum_{j=1}^{n} \lambda^{i_j} p_j f(x_j) \right)$$
$$= \frac{1}{(n+\lambda-1)^k} \sum_{j=1}^{n} \left(\sum_{(i_1,\dots,i_n)\in S_k} \frac{k!}{i_1!\dots i_n!} \lambda^{i_j} \right) p_j f(x_j), \quad k \in \mathbb{N}_+. \tag{15}$$

The multinomial theorem shows that

$$\sum_{(i_1,\dots,i_n)\in S_k} \frac{k!}{i_1!\dots i_n!} \lambda^{i_j} = (n+\lambda-1)^k, \quad 1 \le j \le n,$$

hence (15) implies (14).

The proof of Theorem 3 (a) is based on the following interesting result. The σ-algebra of Borel subsets of $\mathbb{R}^n$ is denoted by $\mathcal{B}^n$.

Lemma 15 *Let $p_1,\dots,p_n$ be a discrete distribution with $n \ge 2$, and let $\lambda > 1$. Let $l \in \{1,\dots,n\}$ be fixed. e_l denotes the vector in $\mathbb{R}^n$ that has 0s in all coordinate positions except the lth, where it has a 1. Let $q_1,\dots,q_n$ be also a discrete distribution such that $q_j > 0$ $(1 \le j \le n)$ and*

$$q_l > \max(q_1, \dots q_{l-1}, q_{l+1}, \dots, q_n). \tag{16}$$

If

$$g : \left\{ (t_1, \dots, t_n) \in \mathbb{R}^n \,|\, t_j > 0\,(1 \le j \le n), \quad \sum_{j=1}^{n} t_j = 1 \right\} \to \mathbb{R}$$

is a bounded function for which

$$\tau_l := \lim_{e_l} g$$

exists in $\mathbb{R}$ and $p_l > 0$, then

$$\lim_{k\to\infty} \sum_{(i_1,\dots,i_n)\in S_k} \frac{k!}{i_1!\dots i_n!} q_1^{i_1}\dots q_n^{i_n} g\left(\frac{\lambda^{i_1}p_1}{\sum_{j=1}^{n}\lambda^{i_j}p_j},\dots,\frac{\lambda^{i_n}p_n}{\sum_{j=1}^{n}\lambda^{i_j}p_j}\right) = \tau_l. \quad (17)$$

Proof. To prove the result, we can obviously suppose that $l = 1$.

For the sake of clarity, we shall denote the element $(i_1,\dots, i_n)$ of S_k by $(i_{1k},\dots, i_{nk})$ $(k \in \mathbb{N}_+)$.

Let $\xi_k := (\xi_{1k},\dots, \xi_{nk})$ $(k \in \mathbb{N}_+)$ be a $(\mathbb{R}^n, \mathcal{B}^n)$-random variable on a probability space $(\Omega, \mathcal{A}, P)$ such that ξ_k has multinomial distribution of order k and with parameters $q_1,\dots, q_n$. A fundamental theorem of the statistics (see [9], Theorem 5.4.13), which is based on the multidimensional central limit theorem and the Cochran-Fisher theorem, implies that

$$\lim_{k\to\infty} P\left(\sum_{j=1}^{n}\frac{(\xi_{jk}-kq_j)^2}{kq_j} < t\right) = F_{n-1}(t), \quad t \in \mathbb{R}, \quad (18)$$

where F_{n-1} means the distribution function of the Chi-squared distribution (χ^2-distribution) with n - 1 degrees of freedom.

Choose $0 < \varepsilon < 1$. Since F_{n-1} is continuous and strictly increasing on $]0, \infty[$, there exists a unique $t_\varepsilon > 0$ such that

$$F_{n-1}(t_\varepsilon) = 1 - \varepsilon.$$

Define

$$S_k^1 := \left\{ (i_{1k},\dots, i_{nk}) \in S_k \middle| \sum_{j=1}^{n} k \frac{\left(\frac{i_{jk}}{k} - q_j\right)^2}{q_j} < t_\varepsilon \right\}.$$

The definition of the set S_k^1 shows that

$$\sum_{(i_{1k},\dots,i_{nk})\in S_k^1} \frac{k!}{i_{1k}!\dots i_{nk}!} q_1^{i_{1k}}\dots q_n^{i_{nk}} = P\left((\xi_{1k},\dots,\xi_{nk}) \in S_k^1\right) \quad (19)$$

$$= P\left(\sum_{j=1}^{n} k\frac{\left(\frac{\xi_{jk}}{k} - q_j\right)^2}{q_j} < t_\varepsilon\right) = P\left(\sum_{j=1}^{n} \frac{(\xi_{jk} - kq_j)^2}{kq_j} < t_\varepsilon\right)$$

$$= F_{n-1}(t_\varepsilon) + \left(P\left(\sum_{j=1}^{n} \frac{(\xi_{jk} - kq_j)^2}{kq_j} < t_\varepsilon\right) - F_{n-1}(t_\varepsilon)\right)$$

$$= 1 - \varepsilon + \delta_\varepsilon(k), \quad k \in \mathbb{N}_+, \tag{20}$$

where by (18)

$$\lim_{k\to\infty} \delta_\varepsilon(k) = 0. \tag{21}$$

For $j = 1,..., n$ construct the sequences $(I_k^j)_{k\geq 1}$ by

$$I_k^j := i_{jk}^*, \text{ if } \left|\frac{i_{jk}^*}{k} - q_j\right| = \max\left\{\left|\frac{i_{jk}}{k} - q_j\right| \,|\, (i_{1k}, \ldots, i_{nk}) \in S_k^1\right\}, \quad k \in \mathbb{N}_+. \tag{22}$$

We claim that

$$\lim_{k\to\infty} \frac{I_k^j}{k} = q_j, \quad 1 \le j \le n. \tag{23}$$

Fix $1 \le j \le n$. If (23) is false, then (22) yields that we can find a positive number ρ, a strictly increasing sequence $(k_u)_{u\geq 1}$and points

$$(i_{1k_u}, \ldots, i_{nk_u}) \in S_{k_u}^1, \quad u \in \mathbb{N}_+ \tag{24}$$

such that

$$\left|\frac{i_{jk_u}}{k_u} - q_j\right| \ge \rho, \quad u \in \mathbb{N}_+,$$

and therefore,

$$k_u \frac{\left(\frac{i_{jk_u}}{k_u} - q_j\right)^2}{q_j} \ge k_u \frac{\rho^2}{q_j} \to \infty \text{ as } u \to \infty,$$

contrary to (24).

Let

$$q := \max(q_2, \ldots, q_n).$$

It follows from (16) that

$$\gamma := \frac{1}{3}(q_1 - q) > 0. \tag{25}$$

By (22) and (23), we can find an integer k_γ such that for each $k > k_\gamma$

$$\left|\frac{i_{jk}}{k} - q_j\right| \le \left|\frac{I_k^j}{k} - q_j\right| < \gamma, \quad (i_{1k}, \ldots, i_{nk}) \in S_k^1, \quad 1 \le j \le n.$$

Thus, for every $k > k_\gamma$

$$\frac{i_{1k}}{k} > q_1 - \gamma \text{ and } \frac{i_{jk}}{k} < q_j + \gamma, \quad 2 \le j \le n, \quad (i_{1k}, \ldots, i_{nk}) \in S_k^1,$$

and hence, we get from (25) that

$$i_{1k} - i_{jk} > k\gamma \quad 2 \le j \le n, \quad (i_{1k}, \ldots, i_{nk}) \in S_k^1, \quad k > k_\gamma. \tag{26}$$

We can see that

$$i_{1k} - i_{jk} \to \infty \text{ as } k \to \infty, \quad 2 \le j \le n, \quad (i_{1k}, \ldots, i_{nk}) \in S_k^1. \tag{27}$$

Now, set $S_k^2 := S_k \backslash S_k^1 (k \in \mathbb{N}_+)$, and consider the sequences

$$a_k^1 := \sum_{(i_{1k}, \ldots, i_{nk}) \in S_k^1} \frac{k!}{i_{1k}! \ldots i_{nk}!} q_1^{i_{1k}} \ldots q_n^{i_{nk}} g\left(\frac{\lambda^{i_{1k}} p_1}{\sum_{j=1}^n \lambda^{i_{jk}} p_j}, \ldots, \frac{\lambda^{i_{nk}} p_n}{\sum_{j=1}^n \lambda^{i_{jk}} p_j}\right),$$

and

$$a_k^2 := \sum_{(i_{1k}, \ldots, i_{nk}) \in S_k^2} \frac{k!}{i_{1k}! \ldots i_{nk}!} q_1^{i_{1k}} \ldots q_n^{i_{nk}} g\left(\frac{\lambda^{i_{1k}} p_1}{\sum_{j=1}^n \lambda^{i_{jk}} p_j}, \ldots, \frac{\lambda^{i_{nk}} p_n}{\sum_{j=1}^n \lambda^{i_{jk}} p_j}\right),$$

where $k \in \mathbb{N}_+$. The sum of these sequences is just the studied sequence in (17).

Since $p_1 > 0$, we obtain from (27) that

$$\lim_{k\to\infty} \frac{\lambda^{i_{1k}} p_1}{\sum_{j=1}^{n} \lambda^{i_{jk}} p_j} = 1, \quad (i_{1k}, \ldots, i_{nk}) \in S_k^1, \tag{28}$$

and

$$\lim_{k\to\infty} \frac{\lambda^{i_{lk}} p_1}{\sum_{j=1}^{n} \lambda^{i_{jk}} p_j} = 0, \quad 2 \le l \le n, \quad (i_{1k}, \ldots, i_{nk}) \in S_k^1. \tag{29}$$

According to (26), the convergence is uniform for all the possible sequences in (28) and (29); hence, for every $\varepsilon_1 > 0$, we can find an integer $k_{\varepsilon_1} > k_\gamma$ that for all $k > k_{\varepsilon_1}$

$$\tau_1 - \varepsilon_1 < g\left(\frac{\lambda^{i_{1k}} p_1}{\sum_{j=1}^{n} \lambda^{i_{jk}} p_j}, \ldots, \frac{\lambda^{i_{nk}} p_n}{\sum_{j=1}^{n} \lambda^{i_{jk}} p_j}\right) < \tau_1 + \varepsilon_1, \quad (i_{1k}, \ldots, i_{nk}) \in S_k^1. \tag{30}$$

Bringing in (19-20), we find that

$$P\left((\xi_{1k}, \ldots, \xi_{nk}) \in S_k^2\right) = \varepsilon - \delta_\varepsilon(k), \quad k \in \mathbb{N}_+,$$

and therefore, thanks to (19-20), (30) and the boundedness of g ($|g| \le m$)

$$\begin{aligned} &(1-\varepsilon + \delta_\varepsilon(k))(\tau_1 - \varepsilon_1) - (\varepsilon - \delta_\varepsilon(k))m \le a_k^1 + a_k^2 \\ &\le (1-\varepsilon + \delta_\varepsilon(k))(\tau_1 + \varepsilon_1) + (\varepsilon - \delta_\varepsilon(k))m, \quad k > k_{\varepsilon_1}. \end{aligned}$$

Consequently, by (21)

$$\begin{aligned} (1-\varepsilon)(\tau_1 - \varepsilon_1) - \varepsilon m &\le \liminf_{k\to\infty}(a_k^1 + a_k^2) \le \limsup_{k\to\infty}(a_k^1 + a_k^2) \\ &\le (1-\varepsilon)(\tau_1 + \varepsilon_1) + \varepsilon m, \end{aligned}$$

and this proves the convergence claim (17).

The proof is now complete. ■

The proof of Theorem 3.

Proof. (a) We have only to observe that for every fixed $1 \le l \le n$

$$\lim_{k\to\infty} \frac{1}{(n+\lambda-1)^k} \sum_{(i_1,\ldots,i_n)\in S_k} \frac{k!}{i_1!\ldots i_n!} \lambda^{i_l} p_l f\left(\frac{\sum_{j=1}^{n} \lambda^{i_j} p_j x_j}{\sum_{j=1}^{n} \lambda^{i_j} p_j}\right) = p_l f(x_l). \tag{31}$$

The case $p_l = 0$ is trivial.

To prove the case $p_l > 0$, define the function

$$g : \left\{ (t_1, \ldots, t_n) \in \mathbb{R}^n | t_j > 0\, (1 \le j \le n),\quad \sum_{j=1}^{n} t_j = 1 \right\} \to \mathbb{R}$$

by

$$g(t_1, \ldots, t_n) := f\left(\sum_{j=1}^{n} t_j x_j\right).$$

Consequently, the limit in (31) can be written in the form

$$\lim_{k\to\infty} p_l \sum_{(i_1,\ldots,i_n)\in S_k} \frac{k!}{i_1!\ldots i_n!} \left(\frac{1}{n+\lambda-1}\right)^{i_1} \cdots \left(\frac{1}{n+\lambda-1}\right)^{i_{l-1}} \left(\frac{\lambda}{n+\lambda-1}\right)^{i_l}$$
$$\cdot \left(\frac{1}{n+\lambda-1}\right)^{i_{l+1}} \cdots \left(\frac{1}{n+\lambda-1}\right)^{i_n} g\left(\frac{\lambda^{i_1} p_1}{\sum_{j=1}^{n} \lambda^{i_j} p_j}, \ldots, \frac{\lambda^{i_n} p_n}{\sum_{j=1}^{n} \lambda^{i_j} p_j}\right).$$

Now, we can apply Lemma 15 with

$$q_j = \frac{1}{n+\lambda-1}, \quad 1 \le j \le n, \quad j \ne l, \text{ and } q_l = \frac{\lambda}{n+\lambda-1}$$

and

$$\lim_{c_l} g = f(x_l), \quad 1 \le l \le n.$$

(b) Elementary considerations show this part of the theorem.

The proof is complete. ■

The proof of Theorem 5.

Proof. Theorem A confirms that *f* is bounded on the set

$$G := \left\{ \sum_{j=1}^{n} t_j x_j \in C | t_j \ge 0 \quad (1 \le j \le n), \quad \sum_{j=1}^{n} t_j = 1 \right\},$$

where t_j $(1 \le j \le n)$ is also rational if *f* is mid-convex.

It is elementary that for every $(i_1,..., i_n) \in S_k$

$$\lim_{\lambda \to \infty} \frac{\lambda^{i_l}}{(n + \lambda - 1)^k} = \begin{cases} 1, & \text{if } i_l = k \\ 0, & \text{if } i_l < k \end{cases}, \quad 1 \le l \le n.$$

By the definition of the set S_k, (0,..., 0, k, 0,..., 0) (the vector has 0s in all coordinate positions except the lth) is the only element of S_k for which $i_l = k$ $(1 \le l \le n)$. By using the boundedness of *f* on *G*, the previous assumptions imply the result, bringing the proof to an end. ■

ACKNOWLEDGEMENTS

This study was supported by the Hungarian National Foundations for Scientific Research Grant No. K73274.

COMPETING INTERESTS

The author declares that he has no competing interests.

REFERENCES

1. Hardy GH, Littlewood JE, Pólya G: *Inequalities. Cambridge Mathematical Library Series.* Cambridge University Press; 1967.

2. Pečarić JE, Volenec V: **Interpolation of the Jensen inequality with some applications, Österreích.** *Akad Wiss Math-Natur KI Sitzungsber II* 1988, **197:** 463–467.

3. Pečarić JE, Svrtan D: **Unified approach to refinements of Jensen's inequalities.** *Math Inequal Appl* 2002, **5:** 45–47.

4. Horváth L, Pečarić JE: **A refinement of the discrete Jensen's inequality.** *Math Inequal Appl* (to appear)

5. Horváth L: **A method to refine the discrete Jensen's inequality for convex and mid-convex functions.** *Mathematical and Computer Modelling,* in press. 10.1016/j.mcm.2011.05.060

6. Horváth L: **Inequalities corresponding to the classical Jensen's inequality.** *J Math Inequal* 2009, **3:** 189–200.

7. Xiao Z-G, Srivastava HM, Zhang Z-H: **Further refinements of the Jensen inequalities based upon samples with repetitions.** *Math Comput Mod* 2010, **51:** 592–600.

8. Mitrinović DS, Pečarić JE, Fink AM: *Classical and New Inequalities in Analysis, vol. 61 of Mathematics and Its Applications.* Kluwer Academic Publishers, Dordrecht; 1993.

9. Dacunha-Castello D, Duflo M: *Probability and Statistics I.* Springer-Verlag, New York; 1986.

5

FIXED-POINT ITERATION METHOD FOR SOLVING THE CONVEX QUADRATIC PROGRAMMING WITH MIXED CONSTRAINTS

Ruopeng Wang, Hong Shi, Kai Ruan, Xiangyu Gao

Department of Mathematics & Physics, Beijing Institute of Petrochemical Technology, Beijing, China

ABSTRACT

The present paper is devoted to a novel smoothing function method for convex quadratic programming problem with mixed constrains, which has important application in mechanics and engineering science. The problem is reformulated as a system of non-smooth equations, and then a smoothing function for the system of non-smooth equations is proposed. The condition of convergences of this iteration algorithm is given. Theory analysis and primary numerical results illustrate that this method is feasible and effective.

KEYWORDS

Fixed-Point Iteration; Convex Quadratic Programming Problem; Convergence; Smoothing Function

1. INTRODUCTION

Consider the following convex quadratic programming problem

$$\min \frac{1}{2} x^{\mathrm{T}} Q x + c^{\mathrm{T}} x$$

$$\text{s.t.} \begin{cases} A_{11} x_I + A_{12} x_{II} \geq b_I \\ A_{21} x_I + A_{22} x_{II} = b_{II} \end{cases} \quad (1)$$

where $Q = \begin{bmatrix} Q_{11} & Q_{12} \\ Q_{21} & Q_{22} \end{bmatrix}$ is an $n \times n$ symmetrical positive semi-definite partitioned matrix, $A = \begin{bmatrix} A_{11} & A_{12} \\ A_{21} & A_{22} \end{bmatrix}$ is an $m \times n$ partitioned matrix, $b = \begin{bmatrix} b_I \\ b_{II} \end{bmatrix}$ and $c = \begin{bmatrix} c_I \\ c_{II} \end{bmatrix}$ are vectors in R^m and R^n, respectively. And Q_{ij}, $A_{ij} \, (i, j = 1, 2)$ are the partitioned matrix of Q and A respectively.

As is well known, this problem models a large spectrum of applications in computer science. Operation research and engineering as the nonlinear programming is solved eventually based on the convex quadratic programming problem. Therefore, the study on its effective solvers becomes a prolonged research subject.

Common effective algorithm is proposed related to certain schemes of convex quadratic programming problem, see [1-5]. Moreover, these methods are not polynomial algorithm in [6]. And the theory and algorithm of quadratic programming are presented completely. According to the optimization condition (KKT) and the duality theory of quadratic programming, the fixed-point iteration method is obtained for equality constrained (see [7]) and inequality constrained (see [8]) convex quadratic programming problem respectively. In fact, equality and inequality can be converted with inputting artificial variables and slack variables. The main default is to increase the scale and numerical difficulty while introduces these variables. Motivated by smoothing methods of [9-11] and fixed-point method of [7,8], the paper mainly concerns about the mixed constraint quadratic programming and the fixed-point iteration method is given. And also the rank of the coefficient matrix is not full. The method considering is fulfilled efficiently. Thus the proposing question in [8] is solved.

2. SCHEME OF ALGORITHM

According to the dual theory, the dual problem of primal problem (1) is

$$\max -\frac{1}{2}x^T Qx + b^T y$$

$$\text{s.t.} \begin{cases} Q_{11}x_I + Q_{12}x_{II} - A_{11}^T y_I - A_{21}^T y_{II} + c_I \geq 0 \\ Q_{21}x_I + Q_{22}x_{II} - A_{21}^T y_I - A_{22}^T y_{II} + c_{II} = 0 \end{cases} \tag{2}$$

where $y = \begin{bmatrix} y_I \\ y_{II} \end{bmatrix} \in R^m$.

Let $u = \begin{bmatrix} u_I \\ u_{II} \end{bmatrix}$, $u_I = \begin{bmatrix} x_I \\ y_I \end{bmatrix}$, $u_{II} = \begin{bmatrix} x_{II} \\ y_{II} \end{bmatrix}$, we know if $x^* = \begin{bmatrix} x_I^* \\ x_{II}^* \end{bmatrix} \in R^n$ is the optimal solution of (1),there exists a vector $y^* = \begin{bmatrix} y_I^* \\ y_{II}^* \end{bmatrix} \in R^m$, an $n \times n$ symmetrical positive semi-definite partitioned matrix, and (x^*, y^*) is the optimal solution of (2).

Let us define the following notation

$$F = \left\{ u = (x_I, y_I, x_{II}, y_{II}) \in R^{n+m} \middle| u_I \geq 0 \right\}$$

$F^* = \{ u^* = (x_I^*, y_I^*, x_{II}^*, y_{II}^*) | (x_I^*, x_{II}^*)$ is the optimal solution of (1) and $(x_I^*, y_I^*, x_{II}^*, y_{II}^*)$ is the optimal solution of (2) }.

It is easy to verify that F is a close set.

Suppose that the feasible region $F^{\bullet}$ is not empty. A necessary and sufficient condition for a optimal solution is that:

1) Primal feasibility

$$A_{11}x_I + A_{12}x_{II} - b_I \geq 0$$

$$A_{21}x_I + A_{22}x_{II} - b_{II} = 0$$

2) Dual feasibility

$$Q_{11}x_I + Q_{12}x_{II} - A_{11}^{\mathrm{T}}y_I - A_{21}^{\mathrm{T}}y_{II} + c_I \geq 0$$
$$Q_{21}x_I + Q_{22}x_{II} - A_{21}^{\mathrm{T}}y_I - A_{22}^{\mathrm{T}}y_{II} + c_{II} = 0$$

3) Complementary condition

$$x_I^{\mathrm{T}}\left[Q_{11}x_I + Q_{12}x_{II} - A_{11}^{\mathrm{T}}y_I - A_{21}^{\mathrm{T}}y_{II} + c_I\right] = 0$$

$$y_I^{\mathrm{T}}\left[A_{11}x_I + A_{12}x_{II} - b_I\right] = 0$$

The scheme are rewritten below

$$(Mu+q)_{II} = 0\ u_I \geq 0, (Mu+q)_I \geq 0, u^{\mathrm{T}}(Mu+q)_I = 0 \tag{3}$$

Where

$$M = \begin{bmatrix} Q_{11} & -A_{11}^{\mathrm{T}} & Q_{12} & -A_{21}^{\mathrm{T}} \\ A_{11} & 0 & A_{12} & 0 \\ Q_{21} & -A_{12}^{\mathrm{T}} & Q_{22} & -A_{22}^{\mathrm{T}} \\ A_{21} & 0 & A_{22} & 0 \end{bmatrix}$$

$$q = \begin{bmatrix} q_I \\ q_{II} \end{bmatrix}, q_I = \begin{bmatrix} c_I \\ -b_I \end{bmatrix}, q_{II} = \begin{bmatrix} c_{II} \\ -b_{II} \end{bmatrix}$$

More formally, the following link fixed-point iteration problem is performed

$$u = \max\left(0, u - r\left(Mu + q\right)\right) \tag{4}$$

The component form is

$$u_j = \max\left(0, u - r\left(M_j u_j + q_j\right)\right) \tag{5}$$

Where $r > 0$.

The question (5) is non-smooth optimization problem, and the Newton-type methods cannot be applied directly. In the paper, we converted the problem into a smooth optimization.

At the beginning, we introduce a function mapping R into R^+, and have the following properties.

Property 1 Let $\theta : R \to R^+$ be a real function, and such that:

1) $\theta(t)$ is strictly convex and differentiable;

2) $\left.\frac{d\theta}{dt}\right|_{t=0} > 0$;

3) $\lim_{t \to -\infty} \theta(t) = 0$;

4) $\lim_{t \to +\infty} \frac{\theta(t)}{t} = 1.$

We assumed that $\theta(t)$ is a convex and differentiable function, which indicate one can apply the classical Newton-type algorithm directly. And the assumption $\theta'(0) > 0$ means that 0 is not a sub-gradient of θ at 0. The hypothesis (3) and (4) indicate the approximate function $\theta(t)$ has the same numerical properties as the max-function.

We consider the following function, and $\theta(t)$ describe as above

$$\theta_p(t) = p\theta\left(\frac{t}{p}\right) \tag{6}$$

For any $p \in (0,1]$, $\theta_p(t)$ satisfies the following properties:

Theorem 1 Let be a function given as (6), then for any arbitrary $p \in (0,1]$, $\theta_p(t)$ is strictly convex and differentiable, and we have $\left.\frac{\mathrm{d}\theta_p(t)}{\mathrm{d}t}\right|_{t=0} > 0$.

Proof According to property 1 and (6), the convexity and differentiability are straightforward.

Nothing that

$$\left.\frac{\mathrm{d}\theta_p(t)}{\mathrm{d}t}\right|_{t=0} = \theta'\left(\frac{t}{p}\right) \geq \theta'(t)$$

The property (2) implies $\left.\frac{\mathrm{d}\theta_p(t)}{\mathrm{d}t}\right|_{t=0} > 0$.

Theorem 2 Let $\theta_p(t): R \to R^+$ be a function given as (6), then $\lim_{p\to+0} \theta_p(t) = \max\{0,t\}$. That is, the function $\theta_p(t)$ uniformly converges to the max-function, when the adjustable parameter p approaches 0.

Proof For t fixed, we can compute

$$\lim_{p\to+0} \theta_p(t) = \lim_{p\to+0} p\theta\left(\frac{t}{p}\right)$$

If $t<0$, then $\lim_{p\to+0} p\theta\left(\frac{t}{p}\right)=0$; otherwise, change variable $z\equiv\frac{t}{p}$,

$$\lim_{p\to+0}\theta_p(t)=\lim_{z\to+\infty} t\cdot\frac{\theta(z)}{x}=t$$

Hence, $\lim_{p\to+0}\theta_p(t)=\max\{0,t\}$ for all t.

From the discussion above, we introduce the following function in the paper

$$\theta(t)=\ln\left(1+e^t\right)$$

and

$$\theta_p(t)=p\theta\left(\frac{t}{p}\right)=p\ln\left(1+e^{\frac{t}{p}}\right)$$

It is easily known that satisfies the properties (1) to (4), together with theorem 1 and 2. Thus, the max-function in (3) replaced by $\theta_p(t)=p\ln\left(1+e^{\frac{t}{p}}\right)$, and the fixed-point iteration which basing on smoothing function is obtained.

Algorithm

Step 1 Set a initial point $u^0\in R^{m+n}$, choose $\varepsilon>0$, $0<p_0\le1$, $l\in(0,1)$, let $k=0$;

Step 2 Apply fixed-point algorithms to solve

$$u_j^{k+1}=\theta_{i,p}\left(u_j^k-r_k\left(M_{,j}u_j^k+q_j\right)\right)$$

$$j=1,2,\cdots,n+m \tag{7}$$

Step 3 Terminal condition: $\left|\left(u^k\right)^{\mathrm{T}}\left(M_j u^k + q\right)\right| < \varepsilon$, stop; otherwise, go to Step 4;

Step 4 Let $p_{k+1} = lp_k$, $k = k+1$, go to step 2.

The Jacobian matrix in (7) is shown

$$B = V\left(I - rM^{\mathrm{T}}\right) \tag{8}$$

Where

$$V = \begin{pmatrix} v_1 & & \\ & \ddots & \\ & & v_{n+m} \end{pmatrix}, \quad v_j = \frac{e^{\frac{u_j^k - r_k\left(M_j u_j^k + q_j\right)}{p}}}{1 + e^{\frac{u_j^k - r_k\left(M_j u_j^k + q_j\right)}{p}}}$$

3. CONVERGENCE ANALYSES

Now we discuss the convergence of the algorithm. To begin with, two simple lemmas are introduced that is the basis of our theoretical analysis. Since their proof can be found in reference [12], we here omit the proof due to space.

Lemma 1 [12] A necessary and sufficient condition for the fixed-point iteration method to be convergent is that the spectral radius of Jacobian matrix B satisfies $\rho(B) < 1$, where the spectral radius $\rho(B)$ be defined as

$\rho(B) = \sqrt{\max \lambda_{B^{\mathrm{T}}B}}$, and denotes the eigenvalue of matrix $B^{\mathrm{T}}B$.

Lemma 2 [12] Suppose M is a matrix, whose eigenvalues can be ordered from the smallest to largest, i.e. $|\lambda_1| \le |\lambda_2| \le \cdots \le |\lambda_{n+m}|$, and V is diagonal matrix with nonnegative diagonal elements, then the maximum norm of the matrix VM's eigenvalue such that $|\bar{\lambda}| \le \max_j \{|v_j|\} |\lambda_j|$.

Theorem 3 suppose $\lambda_1, \lambda_2, \cdots, \lambda_{n+m}$ be the eigenvalues of the Jacobian matrix M, if one of the following conditions is true, then we can select the proper parameter r_k makes the algorithm convergent:

(1) The real part of λ_j is positive, that is $\operatorname{Re}\lambda_j > 0 \ (j = 1, 2, \cdots, n+m)$;

(2) All of the eigenvalues $\lambda_j (j = 1, 2, \cdots, n+m)$ of matrix M are real.

Proof Suppose $\mu_j (j = 1, 2, \cdots, n+m)$ are eigenvalues of the Jacobian matrix in (8), and matrix M's eigenvalues denote as $\lambda_j (j = 1, 2, \cdots, n+m)$.

Let $\lambda = x + y\mathrm{i}$ (where $\mathrm{i} = \sqrt{-1}$).

Note that

$$|1 - r\lambda| \le 1 \Leftrightarrow r\left(x^2 + y^2\right) \le 2x$$

If $\lambda = 0$, for all positive number of r; while $\lambda \ne 0$, we know $r \le \dfrac{2x}{x^2 + y^2}$.

1) If $\lambda_j (j = 1, 2, \cdots, n+m)$ is positive, we let $r = \dfrac{2\min_j \{\operatorname{Re}\lambda_j\}}{\max_j \{\operatorname{Re}|\lambda_j|^2\}}$, and have $|1 - r\lambda_j| \le 1$

$(j = 1, 2, \cdots, n+m)$;

2) If all of the eigenvalues $\lambda_j\ (j=1,2,\cdots,n+m)$ of matrix M are real, we let $r=\dfrac{2}{\max_j\{\mathrm{Re}|\lambda_j|\}}$, and have $|1-r\lambda_j|\le 1\ \ (j=1,2,\cdots,n+m)$.

For all discussed above, associate with Lemma 1, we know that the matrix M has the eigenvalues of $\mu_j\ (j=1,2,\cdots,n+m)$, such that

$$|\mu_j|\le \max_j\{|v_j|\}|1-r\lambda_j|\le \max_j\{|v_j|\}<1$$

According to the theorem above, the algorithm is always convergent if the parameter r is chosen properly. Thus the recent relevant results in reference [8] are extended.

4. NUMERICAL EXPERIMENT AND CONCLUSION

In this section, we implement algorithm for solving system of inequalities in MATLAB7.0 in order to demonstrate the behavior of the algorithm. Due to page limit, we omit to test the effectiveness of our method in generating highly quadratic programming problems.

Example 1 Consider

$$\min z = x_1^2+2x_2^2+x_3^2-2x_1x_2+x_3$$

$$\text{s.t.}\quad \begin{cases} x_1+x_2+x_3=4 \\ 2x_1-x_2+x_3=2 \\ x_1-2x_2-x_3\le 4 \\ x_1,x_2,x_3\ge 0 \end{cases}$$

The exact optimal solution is $x^* = \left(\frac{21}{11}, \frac{43}{22}, \frac{3}{22}\right)^{\mathrm{T}}$, and the optimal value of objective function is $z^* = \frac{175}{44}$.

Example 2 Consider

$$\max z = x_1^2 + x_2^2 + x_3^2$$

$$\text{s.t.} \begin{cases} 3x_1 + 2x_2 + x_3 = 6 \\ 2x_1 - 4x_2 + x_3 \le -8 \\ x_1, x_2, x_3 \ge 0 \end{cases}$$

The exact optimal solution is $x^* = \left(\frac{1}{2}, \frac{9}{4}, 0\right)^{\mathrm{T}}$, and the optimal value of objective function is $z^* = \frac{55}{4}$.

The results are shown in Table1

From the table, we can see that the algorithm is very simple and only applies the traditional fixed-point method. The implementation of this algorithm is rather easy.

Example 3 Consider

$$\min \frac{1}{2} x^{\mathrm{T}} Q x + c^{\mathrm{T}} x$$

$$Ax \ge b$$

and

$$Q = \begin{pmatrix} 100.0429 & 0.9681 & 1.0672 & 1.4239 & 0.2477 & 0.7917 \\ 0.9681 & 100.0988 & 1.1956 & 0.7964 & 0.3543 & 0.8045 \\ 1.0672 & 1.1956 & 100.0919 & 0.7258 & 0.3490 & 1.2315 \\ 1.4239 & 0.7964 & 0.7258 & 101.7046 & 1.1653 & 0.8368 \\ 0.2477 & 0.3543 & 0.3490 & 1.1653 & 101.3110 & 1.1122 \\ 0.7917 & 0.8045 & 1.2315 & 0.8368 & 1.1122 & 100.3080 \end{pmatrix}$$

$$A=\begin{pmatrix} 6.3842 & 6.3349 & 9.7925 & 0.8606 & 6.8983 & 4.7373 \\ 5.1643 & 2.8327 & 0.6622 & 0.5958 & 1.4803 & 2.5038 \\ 1.1316 & 3.9409 & 7.2566 & 7.8306 & 1.1021 & 1.6560 \\ 9.2422 & 9.8371 & 7.7039 & 7.6555 & 6.6527 & 4.9634 \end{pmatrix}$$

$$c=\begin{pmatrix} 4.5647 & 0.1850 & 8.2141 & 4.4470 & 6.1543 & 7.9194 \end{pmatrix}^{T}$$

$$b=\begin{pmatrix} -4.6091 & -3.6910 & -0.8813 & -2.0285 \end{pmatrix}^{T}$$

Table 1. Experiments and calculated data.

Example	Optimal Solution	Objective Value
1	$\begin{pmatrix} 1.909091 \\ 1.954546 \\ 0.136364 \end{pmatrix}$	3.977273
2	$\begin{pmatrix} 0.500000 \\ 2.250000 \\ 0.000000 \end{pmatrix}$	13.750000

Table 2. Experiments and calculated data.

Iterations	Time	Objective Value
10	5.2s	0.00000

The exact optimal solution is

$x^{*}=(0\ \ 0\ \ 0\ \ 0\ \ 0\ \ 0)^{T}$ and the optimal value of objective function is $z^{*}=0$.

Appling the method proposed, the iterations consuming time are shown in Table2

ACKNOWLEDGEMENTS

This work is supported by the Beijing Municipal Commission of Education Science and Technology Program (KM201310017006) and the Beijing URT Program (2012J00018).

REFERENCES

1. D. Goldfarb and A. Idinabi, "A Numerical Stable Dual Method for Solving Strictly Convex Quadratic Programs," Mathematical Programming, Vol. 27, No. 2, 1983, pp. 1-33.
2. Y. Ye and E. Tse, "An Extension of Kamarker's Algorithm to Convex Quadratic Programming," Mathematical Programming, Vol. 47, No. 4, 1989, pp. 157-179.
3. J. L. Zhang and X. S. Zhang, "A Predictor-Corrector Method for Convex Quadratic Programming," Journal of System Science and Mathematical Science, Vol. 23, No. 3, 2003, pp. 353-366.
4. R. D. Monteiro, I. Adler and M. G. Resende, "A Polynomial-Time Primal-Dual Affine Scaling Algorithm for Linear and Convex Quadratic Programming and Its Powerseries Extension," Mathematics of Operations Research, Vol. 15, No. 2, 1990, pp. 191-214.
5. M. W. Zhang and C. C. Huang, "A Primal-Dual Infeasible Interior Point Algorithm for Convex Quadratic Programming Problem with Box Constraints," Journal of Engineering Mathematics, Vol. 18, No. 2, 2001, pp. 85-90.

6. Y. X. Yuan and W. Y. Sun, "Optimization Theory and Method," Science Press, 2002, pp. 111-117.

7. R. P. Wang, "Fixed Iterative Method for Solving the Equality Constrained Convex Quadratic Programming Problem," Journal of Beijing Institute of Petro-Chemical Technology, Vol. 16, No. 3, 2008, pp. 64-66.

8. R. P. Wang, "Fixed Iterative Method for Solving the Inequality Constrained Quadratic Programming Problem," Journal of Beijing Institute of Petro-Chemical Technology, Vol. 15, No. 1, 2007, pp. 1-4.

9. X. S. Li, "An Effective Algorithm for Non-differentiable Problem," Science in China (Series A), Vol. 23, No. 3, 1994, pp. 353- 366.

10. G. Q. Chen and Y. Q. Chen, "An Entropy Function Method for Nonlinear Complementary Problems," Acta Scientiarum Naturalium Universitatis NeiMongol, Vol. 31, No. 5, 2000, pp. 447-451.

11. H. W. Tang and L. W. Zhang, "An Entropy Function Method for Nonlinear Programming Problems," Chinese Science Bulletin, Vol. 39, No. 8, 1994, pp. 682-684.

12. Q. Y. Li, Z. Z. Mo and L. Q. Qi, "Numerical Method for System of Nonlinear Equations," Beijing Science Press, Beijing, 1999.

6

ON EQUIVALENCE OF SOME ITERATIONS CONVERGENCE FOR QUASI-CONTRACTION MAPS IN CONVEX METRIC SPACES

Zhiqun XueEmail author, Guiwen Lv and BE Rhoades

ABSTRACT

We show the equivalence of the convergence of Picard and Krasnoselskij, Mann, and Ishikawa iterations for the quasi-contraction mappings in convex metric spaces.

1. INTRODUCTION

Let (E,d) be a complete metric space and $I=[0,1]$. Denote $E^2=E\times E, I^2=I\times I$. A continuous mapping $W: E^2\times I^2\to E$ is said to be a convex structure on E [1] if for all $u,z_1,z_2\in E,\ \lambda_1,\ \lambda_2\in I$ with $\lambda_1+\lambda_2=1$ such that

$$d(u,W(z_1,z_2;\lambda_1,\lambda_2))\le\lambda_1 d(u,z_1)+\lambda_2 d(u,z_2); \tag{1.1}$$

$$W(z_1,z_2;1,0)=z_1, \qquad W(z_1,z_2;0,1)=z_2. \tag{1.2}$$

If (E,d) satisfies the conditions of convex structure, then (E,d) is called convex metric space that is denoted as (E,d,W).

In the following part, we will consider a few iteration sequences in convex metric space (E, d, W). Suppose that T is a self-map of E.

Picard iteration is as follows:

$$\forall p_0 \in E, \quad p_{n+1} = Tp_n = T^{n+1}p_0, \quad n \geq 0. \tag{1.3}$$

Krasnoselskij iteration is as follows:

$$\forall v_0 \in E, \quad v_{n+1} = W(v_n, Tv_n; 1-\lambda, \lambda), \quad n \geq 0, \tag{1.4}$$

where $\lambda \in [0, 1]$.

Mann iteration is as follows:

$$\forall u_0 \in E, \quad u_{n+1} = W(u_n, Tu_n; 1-a_n, a_n), \quad n \geq 0, \tag{1.5}$$

where $a_n \in [0, 1]$.

Ishikawa iteration is as follows:

$$\begin{aligned} &\forall x_0 \in E, \\ x_{n+1} &= W(x_n, Ty_n; 1-a_n, a_n), \quad n \geq 0, \\ y_n &= W(x_n, Tx_n; 1-b_n, b_n), \quad n \geq 0, \end{aligned} \tag{1.6}$$

where $a_n, b_n \in [0, 1]$ for all $n \geq 0$.

A mapping $T: E \to E$ is called contractive if there exists $L \in (0, 1)$ such that

$$d(Tx, Ty) \leq Ld(x, y), \tag{1.7}$$

for all $x, y \in E$.

The map T is called Kannan mapping [2] if there exists $b \in (0, 1/2)$ such that

$$d(Tx, Ty) \le b[d(x, Tx) + d(y, Ty)], \tag{1.8}$$

for all $x, y \in E$.

A similar definition of mapping is due to the work Chatterjea [3] (that is called Chatterjea mapping), if there exists $c \in (0, 1/2)$ such that

$$d(Tx, Ty) \le c[d(x, Ty) + d(y, Tx)], \tag{1.9}$$

for all $x, y \in E$.

Combining above three definitions, Zamfirescu [4] showed the following result.

Theorem 1.1.

Let (E, d) be a complete metric space and $T : E \to E$ a mapping for which there exist the real numbers $a,\ b,$ and c satisfying $a \in (0, 1),\ b, c \in (0, 1/2)$ such that, for any pair $x, y \in E$, at least one of the following conditions holds:

(z1) $d(Tx, Ty) \le a d(x, y)$;

(z2) $d(Tx, Ty) \le b[d(x, Tx) + d(y, Ty)]$;

(z3) $d(Tx, Ty) \le c[d(x, Ty) + d(y, Tx)]$.

Then T has a unique fixed point, and the Picard iteration converges to fixed point. This class mapping is called Zamfirescu mapping.

In 1974, Ćirić [5] introduced one of the most general contraction mappings and obtained that the unique fixed point can be approximated by Picard iteration. This mapping is called quasi-contractive if there exists $\delta \in (0, 1)$ such that

$$d(Tx, Ty) \le \delta \cdot \max\{d(x, y), d(x, Tx), d(y, Ty), d(x, Ty), d(y, Tx)\}, \tag{1.10}$$

for any $x, y \in E$.

Clearly, every quasi-contraction mapping is the most general of above mappings.

Later on, in 1992, Xu [6] proved that Ishikawa iteration can also be used to approximate the fixed points of quasi-contraction mappings in real Banach spaces.

Theorem 1.2.

Let C be any nonempty closed convex subset of a Banach space X and $T: C \to C$ a quasi-contraction mapping. Suppose that $\alpha_n > 0$ for all n and $\Sigma \alpha_n = \infty$. Then the Ishikawa iteration sequence $\{x_n\}$ defined by (1)–(3) converges strongly to the unique fixed point x^* of T.

In this paper, we will show the equivalence of the convergence of Picard and Krasnoselskij, Mann, and Ishikawa iterations for the quasi-contraction mappings in convex metric spaces.

Lemma 1.3.

Let $\{\rho_n\}_{n=0}^{\infty}$ be a nonnegative sequence which satisfies the following inequality

$$\rho_{n+1} \le (1 - \theta_n)\rho_n + \sigma_n, \quad n \ge 0, \tag{1.11}$$

where $\theta_n \in (0, 1), \Sigma_{n=0}^{\infty} \theta_n = \infty$, and $\sigma_n / \theta_n \to 0$ as $n \to \infty$. Then $\rho_n \to 0$ as $n \to \infty$, (see [7]).

2. RESULTS FOR QUASI-CONTRACTION MAPPINGS

Theorem 2.1.

Let (E, d, W) be a convex metric space, $T: E \to E$ a quasi-contraction mapping with $F(T) \neq \varnothing$. Suppose that $\{\rho_n\}_{n=0}^{\infty}, \{v_n\}_{n=0}^{\infty}$ are defined by the iterative processes (1.3) and (1.4), respectively. Then, the following two assertions are equivalent:

(i)Picard iteration (1.3) converges strongly to the unique fixed point $q \in F(T)$;

(ii)Krasnoselskij iteration (1.4) converges strongly to the unique fixed point $q \in F(T)$.

Proof.

First, we show (i) $\Rightarrow$ (ii), that is, $d(p_n, q) \to 0$ as $n \to \infty \Rightarrow d(v_n, q) \to$ as $n \to \infty$.

From (1.3), (1.4), and (1.1), we can get

$$\begin{aligned} d(v_{n+1}, p_{n+1}) &= d(W(v_n, Tv_n; 1-\lambda, \lambda), Tp_n) \\ &\leq (1-\lambda)d(v_n, Tp_n) + \lambda d(Tv_n, Tp_n) \\ &\leq (1-\lambda)d(v_n, p_n) + (1-\lambda)d(p_n, Tp_n) + \lambda d(Tv_n, Tp_n) \\ &\leq (1-\lambda)d(v_n, p_n) + \frac{1-\lambda}{1-\delta}d(p_n, q) + \lambda d(Tv_n, Tp_n). \end{aligned} \tag{2.1}$$

Next, we consider $d(Tv_n, Tp_n)$. Using (1.10) with $x = p_n$, $y = v_n$, to obtain

$$d(Tv_n, Tp_n) \leq \delta \cdot \max\{d(v_n, p_n), d(v_n, Tv_n), d(p_n, Tp_n), d(v_n, Tp_n), d(p_n, Tv_n)\}. \tag{2.2}$$

Set

$$A_n = \{v_i\}_{i=0}^{n} \cup \{p_i\}_{i=0}^{n} \cup \{Tp_i\}_{i=0}^{n} \cup \{Tv_i\}_{i=0}^{n}, \qquad \gamma_n = \text{diam}\{A_n\}. \tag{2.3}$$

Then $\{A_n\}$ is bounded. Without loss of generality, we let $\gamma_n > 0$ for each n. Indeed, we will show this conclusion from the some following cases.

Case 1.

Let $\gamma_n = d(Tp_i, Tv_j)$ for some $0 \leq i, j \leq n$. Then, from (1.10) and the above γ_n, we have

$$\begin{aligned} \gamma_n &= d(Tp_i, Tv_j) \\ &\le \delta \cdot \max\{d(p_i, v_j), d(v_j, Tv_j), d(p_i, Tp_i), d(v_j, Tp_i), d(p_i, Tv_j)\} \\ &\le \delta \gamma_n < \gamma_n, \end{aligned} \tag{2.4}$$

and it leads to a contradiction. Thus, $\gamma_n \neq d(Tp_i, Tv_j)$. Similarity to $\gamma_n = d(Tp_i, Tp_j)$ or $\gamma_n = d(Tv_i, Tv_j)$ is also impossible.

Case 2.

Let $\gamma_n = d(p_i, v_j)$ for some $0 \le i, j \le n$.

(i) If $j = 0$, then $\gamma_n = d(p_i, v_0)$.

(ii) If $j \ge 1,\ i = 0$, then, from (1.4) and (1.1)

$$\begin{aligned} \gamma_n &= d(p_0, v_j) \\ &= d(p_0, W(v_{j-1}, Tv_{j-1}; 1-\lambda, \lambda)) \\ &\le (1-\lambda) d(p_0, v_{j-1}) + \lambda d(p_0, Tv_{j-1}) \\ &\le (1-\lambda) d(p_0, v_{j-1}) + \lambda \gamma_n, \end{aligned} \tag{2.5}$$

that is, $\gamma_n = d(p_0, v_{j-1})$. By induction on j, we can obtain $\gamma_n = d(p_0, v_0)$.

(iii) If $j \ge 1,\ i \ge 1$, from (1.4) and (1.1)

$$\begin{aligned} \gamma_n &= d(p_i, v_j) \\ &= d(p_i, W(v_{j-1}, Tv_{j-1}; 1-\lambda, \lambda)) \\ &\le (1-\lambda) d(p_i, v_{j-1}) + \lambda d(p_i, Tv_{j-1}) \\ &\le (1-\lambda) d(p_i, v_{j-1}) + \lambda \gamma_n, \end{aligned} \tag{2.6}$$

it implies that $\gamma_n = d(p_i, v_{j-1})$. By induction on j, we can get $\gamma_n = d(p_i, v_0)$.

Case 3.

Let $\gamma_n = d(v_i, v_j)$ for some $0 \le i, j \le n$. Without loss of generality, we set $0 \le i < j \le n$. Then, from (1.4), (1.1)

$$\begin{aligned}\gamma_n &= d(v_i, v_j)\\ &\le d(v_i, W(v_{j-1}, Tv_{j-1}; 1-\lambda, \lambda))\\ &\le (1-\lambda)d(v_i, v_{j-1}) + \lambda d(v_i, Tv_{j-1})\\ &\le (1-\lambda)d(v_i, v_{j-1}) + \lambda\gamma_n,\end{aligned} \tag{2.7}$$

it implies that $\gamma_n = d(v_i, v_{j-1})$, and by induction on j, we may get $\gamma_n = d(v_i, v_i) = 0$, which is a contradiction.

Case 4.

Let $\gamma_n = d(v_i, Tp_j)$ for some $0 \le i, j \le n$.

(i) If $i = 0$, then $\gamma_n = d(v_0, Tp_j)$.

(ii) If $i \ge 1$, from (1.4), (1.1), then

$$\begin{aligned}\gamma_n &= d(v_i, Tp_j)\\ &\le d(W(v_{i-1}, Tv_{i-1}; 1-\lambda, \lambda), Tp_j)\\ &\le (1-\lambda)d(v_{i-1}, Tp_j) + \lambda d(Tv_{i-1}, Tp_j)\\ &\le (1-\lambda)d(v_{i-1}, Tp_j)\\ &\quad + \lambda\delta\cdot\max\{d(v_{i-1}, p_j), d(v_{i-1}, Tv_{i-1}), d(p_j, Tp_j), d(v_{i-1}, Tp_j), d(p_j, Tv_{i-1})\}\\ &\le (1-\lambda)d(v_{i-1}, Tp_j) + \lambda\delta\gamma_n\\ &\le (1-\lambda)d(v_{i-1}, Tp_j) + \lambda\gamma_n,\end{aligned} \tag{2.8}$$

it implies that $\gamma_n = d(v_{i-1}, Tp_j)$ and by induction on i, then $\gamma_n = d(v_0, Tp_j)$.

Case 5.

Let $\gamma_n = d(p_i, Tv_j)$ for some $0 \le i, j \le n$.

(i) If $i = 0$, then $\gamma_n = d(p_0, Tv_j)$.

(ii) If $i \ge 1$, then, from (1.3) and (1.10)

$$\begin{aligned}\gamma_n &= d(p_i, Tv_j)\\ &\le d(Tp_{i-1}, Tv_j)\\ &\le \lambda\delta\cdot\max\{d(p_{i-1}, v_j), d(p_{i-1}, Tp_{i-1}), d(v_j, Tv_j), d(p_{i-1}, Tv_j), d(v_j, Tp_{i-1})\}\\ &\le \lambda\delta\gamma_n,\end{aligned} \tag{2.9}$$

this is a contradiction.

Case 6.

let $\gamma_n = d(v_i, Tv_j)$ for some $0 \le i, j \le n$.

(i) If $i = 0$, then $\gamma_n = d(v_0, Tv_j)$.

(ii) If $i \ge 1$, then, from (1.4) and (1.10)

$$\begin{aligned}\gamma_n &= d(v_i, Tv_j)\\ &\le d(W(v_{i-1}, Tv_{i-1}; 1-\lambda, \lambda), Tv_j)\\ &\le (1-\lambda)d(v_{i-1}, Tv_j) + \lambda d(Tv_{i-1}, Tv_j)\\ &\le (1-\lambda)d(v_{i-1}, Tv_j)\\ &\quad +\lambda\delta\cdot\max\{d(v_{i-1}, v_j), d(v_{i-1}, Tv_{i-1}), d(v_j, Tv_j), d(v_{i-1}, Tv_j), d(v_j, Tv_{i-1})\}\\ &\le (1-\lambda)d(v_{i-1}, Tv_j) + \lambda\delta\gamma_n,\end{aligned} \tag{2.10}$$

it implies that $\gamma_n = d(v_0, Tv_j)$.

Case 7.

Let $y_n = d(p_i, p_j)$ for some $0 \leq i, j \leq n$.

(i) If $i = 0, \ \ j > 0$, then $y_n = d(p_0, p_j)$.

(ii) If $i, j \geq 1$, then, from (1.3), (1.10)

$$
\begin{aligned}
y_n &= d(p_i, p_j) \\
&\leq d(Tp_{i-1}, Tp_{j-1}) \\
&\leq \delta \cdot \max\left\{d(p_{i-1}, p_{j-1}), d(p_{i-1}, Tp_{i-1}), d(p_{j-1}, Tp_{j-1}), d(p_{i-1}, Tp_{j-1}), d(p_{j-1}, Tp_{i-1})\right\} \\
&\leq \delta y_n,
\end{aligned}
\tag{2.11}
$$

it is a contradiction.

Case 8.

let $y_n = d(p_i, Tp_j)$ for some $0 \leq i, j \leq n$.

(i) If $i = 0$, then $y_n = d(p_0, Tp_j)$.

(ii) If $i \geq 1$, then, from (1.3) and (1.10)

$$
\begin{aligned}
y_n &= d(p_i, Tp_j) \\
&\leq d(Tp_{i-1}, Tp_j) \\
&\leq \delta \cdot \max\left\{d(p_{i-1}, p_j), d(p_{i-1}, Tp_{i-1}), d(p_j, Tp_j), d(p_{i-1}, Tp_j), d(p_j, Tp_{i-1})\right\} \\
&\leq \delta y_n,
\end{aligned}
\tag{2.12}
$$

which is a contradiction.

Set

$$\begin{aligned}\eta_n &= \max\{\max\{d(p_i, v_0) : 0 < i \le n\}, \max\{d(v_0, Tp_i) : 0 < i \le n\},\\ &\quad \max\{d(v_0, Tv_i) : 0 < i \le n\}, \max\{d(p_0, p_i) : 0 < i \le n\},\\ &\quad \max\{d(p_0, Tv_i) : 0 < i \le n\}, \max\{d(p_0, Tp_i) : 0 < i \le n\}, M\},\end{aligned} \tag{2.13}$$

where $M = \max\{d(p_0, v_0), d(v_0, Tp_0), d(v_0, Tv_0), d(p_0, Tv_0), d(p_0, Tp_0)\}$.

In view of the above cases, then $\gamma_n = \eta_n$, and we obtain that $\{\gamma_n\}$ is bounded.

Indeed, suppose that $\gamma_n = d(p_i, v_0)$ for some $0 < i \le n$. Then,

$$\begin{aligned}\gamma_n &= d(p_i, v_0)\\ &\le d(p_i, Tv_0) + d(Tv_0, v_0)\\ &= d(Tp_{i-1}, Tv_0) + d(Tv_0, v_0)\\ &\le d(Tp_{i-1}, Tv_0) + d(Tv_0, v_0)\\ &\le \delta \cdot \max\{d(p_{i-1}, v_0), d(v_0, Tv_0), d(p_{i-1}, Tp_{i-1}), d(v_0, Tp_{i-1}), d(p_{i-1}, Tv_0)\}\\ &\quad + d(Tv_0, v_0)\\ &\le \delta\gamma_n + d(Tv_0, v_0),\end{aligned} \tag{2.14}$$

which implies that $\gamma_n \le (1/(1-\delta))d(Tv_0, v_0)$. Similarly, if $\gamma_n = d(v_0, Tp_i)$ or $\gamma_n = d(v_0, Tv_i)$, we also obtain $\gamma_n \le (1/(1-\delta))d(Tv_0, v_0)$.

On the other hand, suppose that $\gamma_n = d(p_0, p_i)$ for some $0 < i \le n$. Then,

$$\begin{aligned}\gamma_n &= d(p_0, p_i)\\ &\le d(p_0, Tp_0) + d(Tp_0, Tp_{i-1})\\ &\le d(p_0, Tp_0)\\ &\quad + \delta \cdot \max\{d(p_0, p_{i-1}), d(p_0, Tp_0), d(p_{i-1}, Tp_{i-1}), d(p_0, Tp_{i-1}), d(p_{i-1}, Tp_0)\}\\ &\le d(p_0, Tp_0) + \delta\gamma_n,\end{aligned} \tag{2.15}$$

which implies that $y_n \le (1/(1-\delta))d(Tp_0, p_0)$. Similarly, if $y_n = d(p_0, Tv_i)$ or $y_n = d(p_0, Tp_i)$, we also obtain $y_n \le (1/(1-\delta))d(Tp_0, p_0)$. Therefore, from the above results, we obtain that $y_n \le (1/(1-\delta))M$, that is, $\{A_n\}$ is bounded.

For each $n \in \mathbb{N}$, define

$$B_n = \{v_i\}_{i\ge n} \cup \{p_i\}_{i\ge n} \cup \{Tp_i\}_{i\ge n} \cup \{Tv_i\}_{i\ge n}, \qquad R_n = \operatorname{diam}(B_n). \tag{2.16}$$

Then, using the same proof above, it can be shown that

$$\begin{aligned} R_n = \operatorname{diam}(B_n) = \max\{&\sup\{d(p_i, v_n) : i \ge n\}, \sup\{d(v_n, Tp_i) : i \ge n\}, \\ &\sup\{d(p_n, Tv_i) : i \ge n\}, \sup\{d(v_n, Tv_i) : i \ge n\}, \\ &\sup\{d(p_n, p_i) : i > n\}, \sup\{d(p_n, Tp_i) : i \ge n\}\}. \end{aligned} \tag{2.17}$$

If $R_n = \sup\{d(p_i, v_n) : i \ge n\}$, and using (1.1) and (1.4), then

$$\begin{aligned} R_n &= \sup_{i\ge n} d(p_i, v_n) \\ &= \sup_{i\ge n} d(p_i, W(v_{n-1}, Tv_{n-1}; 1-\lambda, \lambda)) \\ &\le \sup_{i\ge n} \{(1-\lambda)d(p_i, v_{n-1}) + \lambda d(p_i, Tv_{n-1})\} \\ &\le \sup_{i\ge n} \{(1-\lambda)R_{n-1} + \lambda d(Tp_{i-1}, Tv_{n-1})\} \\ &\le \sup_{i\ge n} \{(1-\lambda)R_{n-1} + \lambda\delta \\ &\quad \cdot \max\{d(p_{i-1}, v_{n-1}), d(v_{n-1}, Tv_{n-1}), d(p_{i-1}, Tp_{i-1}), d(v_{n-1}, Tp_{i-1}), d(p_{i-1}, Tv_{n-1})\}\} \\ &\le (1-\lambda)R_{n-1} + \lambda\delta R_{n-1} \\ &= (1-\lambda(1-\delta))R_{n-1} \\ &\le \cdots \\ &\le (1-\lambda(1-\delta))^n R_0 \\ &\to 0 \end{aligned} \tag{2.18}$$

as $n \to \infty$. Since $d(Tv_n, Tp_n) \le R_n$, hence $d(Tv_n, Tp_n) \to 0$ as $n \to \infty$. Similarly, if $R_n = \sup\{d(v_n, Tp_i) : i \ge n\}$ or $R_n = \sup\{d(v_n, Tv_i) : i \ge n\}$,

$R_n = \sup\{d(p_n, Tv_i) : i \geq n\}, R_n = \sup\{d(v_n, Tv_i) : i \geq n\}$, , $R_n = \sup\{d(p_n, p_i) : i > n\}$, $R_n = \sup\{d(p_n, Tp_i) : i \geq n\}$, we may obtain the similar results. Therefore, from (2.1), we get

$$d(v_{n+1}, p_{n+1}) \leq (1-\lambda)d(v_n, p_n) + \sigma_n, \tag{2.19}$$

where $\sigma_n = ((1-\lambda)/(1-\delta))d(p_n, q) + \lambda d(Tv_n, Tp_n)$.

In (2.19), set $\rho_n = d(v_n, p_n)$. Then (2.19) is as follows:

$$\rho_{n+1} \leq (1-\lambda)\rho_n + \sigma_n. \tag{2.20}$$

By Lemma 1.3, we have $d(v_n, p_n) \to 0$ as $n \to \infty$. From the inequality $0 \leq d(v_n, q) \leq d(v_n, p_n) + d(p_n, q)$, we have $\lim_{n\to\infty} d(v_n, q) = 0$.

Conversely, we will prove that (ii) $\Rightarrow$ (i). If $\lambda = 1$, then $v_{n+1} = W(v_n, Tv_n; 0, 1) = Tv_n$ is Picard iteration.

Theorem 2.2.

Let $(E, d, W), T, F(T)$ be as in Theorem 2.1. Suppose that $\{u_n\}_{n=0}^{\infty}, \{x_n\}_{n=0}^{\infty}$ are defined by the iterative processes (1.5) and (1.6), respectively, and $\{a_n\}_{n=0}^{\infty}, \{b_n\}_{n=0}^{\infty}$ are real sequences in $[0, 1]$ such that $\sum_{n=0}^{\infty} a_n = \infty$. Then, the following two assertions are equivalent:

(i) Mann iteration (1.5) converges strongly to the unique fixed point $q \in F(T)$;

(ii) Ishikawa iteration (1.6) converges strongly to the unique fixed point $q \in F(T)$.

Proof.

If the Ishikawa iteration (1.6) converges strongly to q, then setting $b_n = 0$, for all $n \geq 0$, in (1.6), we can get the convergence of Mann iteration (1.5). Conversely, we will show that (i) $\Rightarrow$ (ii). Letting $\lim_{n\to\infty} d(u_n, q) = 0$, we want to prove $\lim_{n\to\infty} d(x_n, q) = 0$.

From (1.5) and (1.6),

$$\begin{aligned}
& d(x_{n+1}, u_{n+1}) \\
& = d(W(x_n, Ty_n; 1-a_n, a_n), W(u_n, Tu_n; 1-a_n, a_n)) \\
& \leq (1-a_n) d(x_n, W(u_n, Tu_n; 1-a_n, a_n)) + a_n d(Ty_n, W(u_n, Tu_n; 1-a_n, a_n)) \\
& \leq (1-a_n)^2 d(x_n, u_n) + a_n(1-a_n) d(x_n, Tu_n) \\
& \quad + (1-a_n) a_n d(Ty_n, u_n) + a_n^2 d(Ty_n, Tu_n) \\
& \leq (1-a_n)^2 d(x_n, u_n) + a_n(1-a_n) d(x_n, u_n) + a_n(1-a_n) d(u_n, Tu_n) \\
& \quad + (1-a_n) a_n d(Ty_n, Tu_n) + (1-a_n) a_n d(Tu_n, u_n) + a_n^2 d(Ty_n, Tu_n) \\
& = (1-a_n) d(x_n, u_n) + 2a_n(1-a_n) d(u_n, Tu_n) + a_n d(Ty_n, Tu_n) \\
& \leq (1-a_n) d(x_n, u_n) + 2a_n(1-a_n) d(u_n, q) \\
& \quad + 2a_n(1-a_n) d(Tu_n, Tq) + a_n d(Ty_n, Tu_n) \\
& \leq (1-a_n) d(x_n, u_n) + 2a_n \frac{1-a_n}{1-\delta} d(u_n, q) + a_n d(Ty_n, Tu_n).
\end{aligned} \tag{2.21}$$

Using (1.10) with $x = y_n, \ y = u_n$, to obtain

$$d(Ty_n, Tu_n) \leq \delta \cdot \max\{d(y_n, u_n), d(u_n, Tu_n), d(y_n, Ty_n), d(y_n, Tu_n), d(u_n, Ty_n)\}, \tag{2.22}$$

set

$$A_{nn} = \{u_i\}_{i=0}^n \cup \{y_i\}_{i=0}^n \cup \{x_i\}_{i=0}^n \cup \{Tu_i\}_{i=0}^n \cup \{Ty_i\}_{i=0}^n \cup \{Tx_i\}_{i=0}^n, \quad \gamma_{nn} = \operatorname{diam}(A_{nn}). \tag{2.23}$$

Applying the similar proof methods of Theorem 2.1, we obtain that $\{A_{nn}\}$ is also bounded. The other proof is the same as that of Theorem 2.1 and is here omitted.

ACKNOWLEDGMENTS

The authors are extremely grateful to Professor B. E. Rhoades of Indiana University for providing useful information and many help. They also thank the referees for their valuable comments and suggestions.

REFERENCES

1. Takahashi W: **A convexity in metric space and nonexpansive mappings. I.** *Kōdai Mathematical Seminar Reports* 1970, **22:** 142–149. 10.2996/kmj/1138846111
2. Kannan R: **Some results on fixed points.** *Bulletin of the Calcutta Mathematical Society* 1968, **60:** 71–76.
3. Chatterjea SK: **Fixed-point theorems.** *Comptes Rendus de l'Académie Bulgare des Sciences* 1972, **25:** 727–730.
4. Zamfirescu T: **Fix point theorems in metric spaces.** *Archiv der Mathematik* 1972, **23:** 292–298.
5. Ćirić LB: **A generalization of Banach's contraction principle.** *Proceedings of the American Mathematical Society* 1974, **45:** 267–273.
6. Xu HK: **A note on the Ishikawa iteration scheme.** *Journal of Mathematical Analysis and Applications* 1992,**167**(2):582–587.
7. Weng X: **Fixed point iteration for local strictly pseudo-contractive mapping.** *Proceedings of the American Mathematical Society*1991,**113**(3):727–731.

7

ESTIMATION OF PARAMETERS OF BOUNDARY VALUE PROBLEMS FOR LINEAR ORDINARY DIFFERENTIAL EQUATIONS WITH UNCERTAIN DATA

Yury Shestopalov[1*], Yury Podlipenko[2], Olexandr Nakonechnyi[2]

[1]Gävle University, Gävle, Sweden

[2]Kyiv National University, Kyiv, Ukraine

ABSTRACT

In this paper we construct optimal, in certain sense, estimates of values of linear functionals on solutions to two-point boundary value problems (BVPs) for systems of linear first-order ordinary differential equations from observations which are linear transformations of the same solutions perturbed by additive random noises. It is assumed here that right-hand sides of equations and boundary data as well as statistical characteristics of random noises in observations are not known and belong to certain given sets in corresponding functional spaces. This leads to the necessity of introducing minimax statement of an estimation problem when optimal estimates are defined as linear, with respect to observations, estimates for which the maximum of mean square error of estimation taken over the above-mentioned sets attains minimal value. Such estimates are called minimax mean square or guaranteed estimates. We establish that the minimax mean square estimates are expressed via solutions of some systems of differential equations of special type and determine estimation errors.

KEYWORDS

Optimal Minimax Mean Square Estimates, Uncertain Data, Two-Point Boundary Value Problems, Random Noises, Observations

1. INTRODUCTION

The theory of finding estimates of solutions to stochastic differential equations has been intensively developing since the classical works of Kalman and Bucy [1] [2]. This theory found numerous applications in the treatment of the results of experiments in physics, biology, medicine, and many other areas of science and technology. Such successful and broad applications are explained by the fact that Kalman-Bucy methods provide differential equations for optimal mean square estimates which can be solved numerically in the real-time mode. At the same time, it should be noted that Krasovskii and Kurzhanskii proposed in [3] [4] an alternative approach to estimating the solutions of differential equations where perturbations and inaccuracies of additional data about solution were not known and the only thing given was that they belong to a certain domain.

Let us formulate a general approach to the problem of minimax estimation. If a state of a system is described by a linear ordinary differential equation

$$\frac{dx(t)}{dt} = Ax(t) + Bv_1(t), \quad x(t_0) = x_0,$$

and a function $y(t)$ is observed in a time interval $[t_0, T]$, where $y(t) = Hx(t) + v_2(t)$, $x(t) \in \mathbb{R}^n$, $y \in \mathbb{R}^m$, $v_2 \in \mathbb{R}^m$, and A, B, H are known matrices, the minimax estimation problem consists in the most accurate determination of a function $x(t)$ at the "worst" realization of unknown quantities $(x_0, v_1(\cdot), v_2(\cdot))$ taken from a certain set. N.N. Krasovskii was the first who stated this problem in [3] . Under different constraints imposed on function $v_2(t)$ and for known function $v_1(t)$ he proposed various methods of estimating inner products $(a, x(T))$, where $a \in \mathbb{R}^n$, in the class of operations linear with respect to observations that minimize the maximal error. Later these estimates were called minimax a priori or guaranteed estimates (see [3] [4]).

This theory was further developed in the works of Chernous'ko, Pshenichnyi, Kuntzevich, Nakonechnyi, Kirichenko, Podilipenko, and their disciples; one may refer e.g. to [4] -[10] and the bibliography therein.

We note that the duality principle elaborated in [3] [4] , and [5] proved its efficiency for the determination of minimax estimates [5] . According to this principle, finding minimax a priori estimates can be reduced to a certain problem of optimal control of a system; this approach enabled one to obtain, under certain restrictions, recurrent equations, namely, the minimax Kalman-Bucy filter (see [5]).

The essential results within the frames of H_∞-theory are obtained in [11] .

In this paper, we study estimation of solutions of boundary value problems (BVPs) for ordinary differential equations at fixed points of interval from additional data about their solutions. Such settings may be considered as inverse problems when additional data are given with errors. We assume that these errors are random with unknown correlation functions. Similar problems arise in data processing of observations of the objects or processes described by BVPs for ordinary differential equations with unknown perturbations of the right-hand sides or boundary conditions. We solve the estimation problems using guaranteed linear estimates that minimize maximal mean square estimation errors. It is shown that optimal guaranteed estimates are expressed via solutions to special BVPs for ordinary differential equations.

2. PRELIMINARIES AND AUXILIARY RESULTS

Assume that it is given a vector-function $f(t)=(f_1(t),f_2(t),\cdots,f_n(t))^{\mathrm{T}}$ with the components belonging to space $L^2(0,T)$ and vectors $f_0=\left(f_1^{(0)},f_2^{(0)},\cdots,f_m^{(0)}\right)^{\mathrm{T}}\in\mathbb{R}^m$ and $f_1=\left(f_1^{(1)},f_2^{(1)},\cdots,f_{n-m}^{(1)}\right)^{\mathrm{T}}\in\mathbb{R}^{n-m}$. Consider the following BVP: find a vector-function $\varphi(t)=(\varphi_1(t),\varphi_2(t),\cdots,\varphi_n(t))^{\mathrm{T}}\in\left(H^1(0,T)\right)^n$ that satisfies a system of linear first-order ordinary differential equations

$$L\varphi(t)\equiv\frac{\mathrm{d}\varphi}{\mathrm{d}t}+A\varphi=f(t),\ t\in(0,T),\ L=\frac{\mathrm{d}}{\mathrm{d}t}+A, \tag{0.1}$$

almost everywhere on an interval $(0,T)$ and the boundary conditions

$$B_0\varphi(0)=f_0,\quad B_1\varphi(T)=f_1 \tag{0.2}$$

at the points 0 and T. Here $A=A(t)$ is an $n\times n$ matrix with the entries $a_{ij}=a_{ij}(t)$ continuous on $[0,T]$;

$$s=\overline{1,n},\ r=\overline{1,m},\ B_0=\left\{b_{rs}^{(0)}\right\},\ \frac{\mathrm{d}\varphi(t)}{\mathrm{d}t}=\left(\frac{\mathrm{d}\varphi_1(t)}{\mathrm{d}t},\frac{\mathrm{d}\varphi_2(t)}{\mathrm{d}t},\cdots,\frac{\mathrm{d}\varphi_n(t)}{\mathrm{d}t}\right)^{\mathrm{T}};$$ and

$B_1=\left\{b_{rs}^{(1)}\right\}$, $r=\overline{1,n-m}$, $s=\overline{1,n}$, are $m\times n$ and $(n-m)\times n$ matrices of rank m and $n-m$, respectively; upper index T denotes transposition of a matrix or a vector and the upper bar throughout the whole text of the paper that e.g. index r

takes all values from 1 to $n-m$; $H^1(a,b)$ is a space of functions absolutely continuous on $[a,b]$ for which the derivative that exists almost everywhere on (a,b) belongs to space $L^2(a,b)$; and

$$\left(H^1(0,T)\right)^n = \overbrace{H^1(0,T)\times\cdots\times H^1(0,T)}^{n \text{ times}}.$$

The problem of finding a function $\varphi(t)$ that satisfies on $(0,T)$ the equation

$$\frac{\mathrm{d}\varphi(t)}{\mathrm{d}t} + A\varphi(t) = 0 \tag{0.3}$$

and the boundary conditions

$$B_0\varphi(0) = 0, \quad B_1\varphi(T) = 0 \tag{0.4}$$

will be called the homogeneous BVP corresponding to BVP (0.1), (0.2).

The solution $\varphi(t)\equiv 0$ to homogeneous BVP (0.3), (0.4) is called the trivial solution.

BVP (0.1), (0.2) can be written in a scalar form:

$$\begin{gathered}
\varphi_1'(t) + a_{11}\varphi_1(t) + a_{12}\varphi_2(t) + \cdots + a_{1n}\varphi_n(t) = f_1(t),\\
\varphi_2'(t) + a_{21}\varphi_1(t) + a_{22}\varphi_2(t) + \cdots + a_{2n}\varphi_n(t) = f_1(t),\\
\vdots\\
\varphi_n'(t) + a_{n1}\varphi_1(t) + a_{n2}\varphi_2(t) + \cdots + a_{nn}\varphi_n(t) = f_n(t),
\end{gathered} \tag{0.5}$$

$$U_i(\varphi) := \sum_{q=1}^{n} b_{iq}^{(0)}\varphi_q(0) = f_i^{(0)}, \ i = \overline{1,m}, \tag{0.6}$$

$$U_{m+i}(\varphi) := \sum_{q=1}^{n} b_{iq}^{(1)} \varphi_q(T) = f_i^{(1)}, \quad i = \overline{1, n-m}.$$

Let

$$\varphi^{(i)}(t) = \left(\varphi_1^{(i)}(t), \varphi_2^{(i)}(t), \cdots, \varphi_n^{(i)}(t)\right)^{\mathrm{T}}, \quad i = \overline{1, n}, \tag{0.7}$$

be a fundamental system of solutions to (0.3) (for the definition, see e.g. [12] p. 179). Then the solutions to (0.3), (0.4) have the form

$$\varphi(t) = c_1 \varphi^{(1)}(t) + c_2 \varphi^{(2)}(t) + \cdots + c_n \varphi^{(n)}(t),$$

where, by virtue of (0.4), constants $c_1, c_2, \cdots, c_n$ must be such that

$$\begin{aligned} &c_1 U_1\left(\varphi^{(1)}\right) + c_2 U_1\left(\varphi^{(2)}\right) + \cdots + c_n U_1\left(\varphi^{(n)}\right) = 0, \\ &c_1 U_2\left(\varphi^{(1)}\right) + c_2 U_2\left(\varphi^{(2)}\right) + \cdots + c_n U_2\left(\varphi^{(n)}\right) = 0, \\ &\qquad\vdots \\ &c_1 U_n\left(\varphi^{(1)}\right) + c_2 U_n\left(\varphi^{(2)}\right) + \cdots + c_n U_n\left(\varphi^{(n)}\right) = 0. \end{aligned} \tag{0.8}$$

Thus, if the matrix

$$\begin{pmatrix} U_1\left(\varphi^{(1)}\right) & U_1\left(\varphi^{(2)}\right) & \cdots & U_1\left(\varphi^{(n)}\right) \\ U_2\left(\varphi^{(1)}\right) & U_2\left(\varphi^{(2)}\right) & \cdots & U_2\left(\varphi^{(n)}\right) \\ \vdots & \vdots & \ddots & \cdots \\ U_n\left(\varphi^{(1)}\right) & U_n\left(\varphi^{(2)}\right) & \cdots & U_n\left(\varphi^{(n)}\right) \end{pmatrix} \tag{0.9}$$

has rank n, the homogeneous BVP has only the trivial solution. The inverse statement is also valid: if the homogeneous BVP has only the trivial solution then the rank of matrix (0.9) equals n (see, for example [13]).

Assume in what follows that homogeneous BVP (0.3), (0.4) corresponding to BVP (0.1), (0.2) has only the trivial solution or what is the same that matrix (0.9) has rank n. As is known [13] , under this assumption, initial BVP (0.1), (0.2) is uniquely solvable at any right-hand sides $f(t)=(f_1(t),f_2(t),\cdots,f_n(t))^{\mathrm{T}}$, $f_0=\left(f_1^{(0)},f_2^{(0)},\cdots,f_m^{(0)}\right)^{\mathrm{T}}\in\mathbb{R}^m$, and $f_1=\left(f_1^{(1)},f_2^{(1)},\cdots,f_{n-m}^{(1)}\right)^{\mathrm{T}}\in\mathbb{R}^{n-m}$.

Formulate the notion of a BVP conjugate to (0.1), (0.2). To this end, introduce the following designations: E_k is the $k\times k$ unit matrix; $O_{k,r}$ is the $k\times r$ null matrix; $k=\overline{1,m}$, $r=\overline{1,m}$, $B_{01}=\left\{b_{ri_k}^{(0)}\right\}$, is a square nondegenerate $m\times m$ submatrix of the matrix $B_0=\left\{b_{rs}^{(0)}\right\}$, $r=\overline{1,m}$, $s=\overline{1,n}$; $l=\overline{1,n-m}$, $r=\overline{1,m}$, $B_{02}=\left\{b_{rj_l}^{(0)}\right\}$, is an $m\times(n-m)$ submatrix of B_0 obtained as a result of deleting in B_0 all columns of matrix B_{01} (so that $\{j_1,\cdots,j_{n-m}\}=\{1,\cdots,n\}\setminus\{i_1,\cdots,i_m\}$); $\hat{B}_0=\left(-B_{02}^{\mathrm{T}}\left(B_{01}^{\mathrm{T}}\right)^{-1},E_{n-m}\right)$ is an $(n-m)\times n$ matrix such that its i_k th column equals k th column of matrix $-B_{02}^{\mathrm{T}}\left(B_{01}^{\mathrm{T}}\right)^{-1}$ (its size is $(n-m)\times n$), $k=\overline{1,m}$, and j_l th column equals lth column of matrix E_{n-m}, $l=\overline{1,n-m}$; $\bar{B}_0=\left(\left(B_{01}^{\mathrm{T}}\right)^{-1},O_{m,n-m}\right)$ is an $m\times n$ matrix such that its i_k th column equals k-th column of matrix $\left(B_{01}^{\mathrm{T}}\right)^{-1}$, $k=\overline{1,m}$, and j_l th column equals lth column of matrix $O_{m,n-m}$, $l=\overline{1,n-m}$; $\tilde{B}_0=\left(O_{n-m,m},E_{n-m}\right)$ is an $(n-m)\times n$ matrix such that its i_k th column equals kth column of matrix $O_{n-m,m}$, $k=\overline{1,m}$, and j_l th column equals lth column of matrix E_{n-m}, $l=\overline{1,n-m}$.

Introduce more similar notations: $k=\overline{1,n-m},\ r=\overline{1,n-m},\ B_{11}=\left\{b_{ri'_k}^{(1)}\right\}$, is a square nondegenerate $(n-m)\times(n-m)$ submatrix of the matrix $B_1=\left\{b_{rs}^{(1)}\right\},\ r=\overline{1,n-m},\ s=\overline{1,n};\ B_{12}=\left\{b_{rj'_l}^{(1)}\right\},\ r=\overline{1,n-m},\quad l=\overline{1,m},$ is a $(n-m)\times m$ submatrix of the matrix B_1 obtained as a result of deleting in B_1 all columns of matrix B_{11} (so that $\{j'_1,\cdots,j'_m\}=\{1,\cdots,n\}\setminus\{i'_1,\cdots,i'_{n-m}\}$); $\hat{B}_1=\left(-B_{12}^{\mathrm{T}}\left(B_{11}^{\mathrm{T}}\right)^{-1},E_m\right)$ is an $m\times n$ matrix such that its i'_k th column equals kth column of matrix $-B_{12}^{\mathrm{T}}\left(B_{11}^{\mathrm{T}}\right)^{-1}$ (the size of the latter is $m\times(n-m)$), $k=\overline{1,n-m}$, and j'_l th column equals lth column of matrix $E_m,\ l=\overline{1,m};$ $\bar{B}_1=\left(\left(B_{11}^{\mathrm{T}}\right)^{-1},O_{n-m,m}\right)$ is an $(n-m)\times n$ matrix such that its i'_k th column equals kth column of matrix $\left(B_{11}^{\mathrm{T}}\right)^{-1},\ k=\overline{1,n-m},$ and j'_l th column equals lth column of matrix $O_{n-m,m},\ l=\overline{1,m};\ \tilde{B}_1=\left(O_{m,n-m},E_m\right)$ is an $m\times n$ matrix such that its i'_k th column equals kth column of matrix $O_{m,n-m}$, $k=\overline{1,n-m},$ and j'_l th column equals lth column of matrix $E_m,\ l=\overline{1,m}.$

By

$$(u,v)_N=\sum_{i=1}^{N}u_iv_i$$

we will denote the inner product of vectors $u=\left(u_1,\cdots,u_N\right)^{\mathrm{T}}$ and $v=\left(v_1,\cdots,v_N\right)^{\mathrm{T}}$ in the Euclidean space $\mathbb{R}^N$.

Then, if $\psi(t)=\left(\psi_1(t),\cdots,\psi_n(t)\right)^{\mathrm{T}}\in\left(H^1(0,T)\right)^n$ we have

$$
\begin{aligned}
&\int_0^T \left(L\varphi(t),\psi(t)\right)_n \mathrm{d}t = \int_0^T \left(\frac{\mathrm{d}\varphi(t)}{\mathrm{d}t} + A\varphi(t),\psi(t)\right)_n \mathrm{d}t \\
&= \int_0^T \left(\frac{\mathrm{d}\varphi(t)}{\mathrm{d}t},\psi(t)\right)_n \mathrm{d}t + \int_0^T \left(A\varphi(t),\psi(t)\right)_n \mathrm{d}t \\
&= \int_0^T \sum_{i=1}^{n} \frac{\mathrm{d}\varphi_i(t)}{\mathrm{d}t}\psi_i(t)\mathrm{d}t + \int_0^T \sum_{i=1}^{n}\sum_{j=1}^{n} a_{ij}\varphi_j(t)\psi_i(t)\mathrm{d}t \\
&= \sum_{i=1}^{n}\varphi_i(t)\psi_i(t)\Big|_0^T - \int_0^T \sum_{i=1}^{n} \frac{\mathrm{d}\psi_i(t)}{\mathrm{d}t}\varphi_i(t)\mathrm{d}t + \int_0^T \sum_{j=1}^{n}\left(\sum_{i=1}^{n} a_{ij}\psi_i(t)\varphi_j(t)\right)\mathrm{d}t \\
&= \left(\varphi(T),\psi(T)\right)_n - \left(\varphi(0),\psi(0)\right)_n - \int_0^T \left(\frac{\mathrm{d}\psi(t)}{\mathrm{d}t},\varphi(t)\right)_n \mathrm{d}t + \int_0^T \sum_{i=1}^{n}\left(\sum_{j=1}^{n} a_{ji}\psi_j(t)\varphi_i(t)\right)\mathrm{d}t \\
&= \left(\varphi(T),\psi(T)\right)_n - \left(\varphi(0),\psi(0)\right)_n + \int_0^T \left(-\frac{\mathrm{d}\psi(t)}{\mathrm{d}t},\varphi(t)\right)_n \mathrm{d}t + \int_0^T \left(\varphi(t),A^{\mathrm{T}}\psi(t)\right)_n \mathrm{d}t \\
&= \left(\varphi(T),\psi(T)\right)_n - \left(\varphi(0),\psi(0)\right)_n + \int_0^T \left(\varphi(t),L^{*}\psi(t)\right)_n,
\end{aligned}
\tag{0.10}
$$

where the differential operator

$$
L^{*} = -\frac{\mathrm{d}}{\mathrm{d}t} + A^{\mathrm{T}}
$$

will be called formally conjugate to operator L.

Let us show that the term $\left(\varphi(T),\psi(T)\right)_n - \left(\varphi(0),\psi(0)\right)_n$ in (0.10) can be represented as

$$
\begin{aligned}
\left(\psi(T),\varphi(T)\right)_n - \left(\psi(0),\varphi(0)\right)_n &= \left(\bar{B}_1\psi(T),B_1\varphi(T)\right)_{n-m} + \left(\hat{B}_1\psi(T),\tilde{B}_1\varphi(T)\right)_m \\
&\quad - \left(\bar{B}_0\psi(0),B_0\varphi(0)\right)_m - \left(\hat{B}_0\psi(0),\tilde{B}_0\varphi(0)\right)_{n-m}.
\end{aligned}
\tag{0.11}
$$

Note first that

$$
B_0\varphi(0) = \begin{pmatrix} \sum_{q=1}^{n} b_{1q}^{(0)}\varphi_q(0) \\ \vdots \\ \sum_{q=1}^{n} b_{nq}^{(0)}\varphi_q(0) \end{pmatrix} = \begin{pmatrix} \sum_{k=1}^{n} b_{1i_k}^{(0)}\varphi_{i_k}(0) + \sum_{l=1}^{n-m} b_{1j_l}^{(0)}\varphi_{j_l}(0) \\ \vdots \\ \sum_{k=1}^{n} b_{mi_k}^{(0)}\varphi_{i_k}(0) + \sum_{l=1}^{n-m} b_{mj_l}^{(0)}\varphi_{j_l}(0) \end{pmatrix} = B_{01}\varphi_1^{(0)}(0) + B_{02}\varphi_2^{(0)}(0),
$$

where

$$\varphi_1^{(0)}(0) := \begin{pmatrix} \varphi_{i_1}(0) \\ \vdots \\ \varphi_{i_m}(0) \end{pmatrix}, \quad \varphi_2^{(0)}(0) := \begin{pmatrix} \varphi_{j_1}(0) \\ \vdots \\ \varphi_{j_{n-m}}(0) \end{pmatrix}.$$

Then $\varphi_1^{(0)}(0) = B_{01}^{-1}B_0\varphi(0) - B_{01}^{-1}B_{02}\varphi_2^{(0)}(0)$, and

$$\begin{aligned}
&(\psi(0),\varphi(0))_n = \left(\psi_1^{(0)}(0),\varphi_1^{(0)}(0)\right)_m + \left(\psi_2^{(0)}(0),\varphi_2^{(0)}(0)\right)_{n-m} \\
&= \left(\psi_1^{(0)}(0), B_{01}^{-1}B_0\varphi(0)\right)_m - \left(\psi_1^{(0)}(0), B_{01}^{-1}B_{02}\varphi_2^{(0)}(0)\right)_m + \left(\psi_2^{(0)}(0),\varphi_2^{(0)}(0)\right)_{n-m} \\
&= \left(\bar{B}_0\psi(0), B_0\varphi(0)\right)_m - \left(B_{02}^{\mathrm{T}}\left(B_{01}^{\mathrm{T}}\right)^{-1}\psi_1^{(0)}(0),\varphi_2^{(0)}(0)\right)_{n-m} + \left(\psi_2^{(0)}(0),\varphi_2^{(0)}(0)\right)_{n-m} \\
&= \left(\bar{B}_0\psi(0), B_0\varphi(0)\right)_m + \left(\left(-B_{02}^{\mathrm{T}}\left(B_{01}^{\mathrm{T}}\right)^{-1}, O_{n-m,n-m}\right)\psi(0),\varphi_2^{(0)}(0)\right)_{n-m} + \left(\left(O_{n-m,n-m}, E_{n-m}\right)\psi(0),\varphi_2^{(0)}(0)\right)_{n-m},
\end{aligned}$$

where $\psi_1^{(0)}(0)$ and $\psi_2^{(0)}(0)$ are vectors composed of components of vector $\psi(0)$ with the numbers equal to the numbers of components of vectors $\varphi_1^{(0)}(0)$ and $\varphi_2^{(0)}(0)$, respectively. Taking into account that

$$\left(-B_{02}^{\mathrm{T}}\left(B_{01}^{\mathrm{T}}\right)^{-1}, O_{n-m,n-m}\right) + \left(O_{n-m,m}, E_{n-m}\right) = \left(-B_{02}^{\mathrm{T}}\left(B_{01}^{\mathrm{T}}\right)^{-1}, E_{n-m}\right) = \hat{B}_0,$$

we have

$$(\psi(0),\varphi(0))_n = \left(\bar{B}_0\psi(0), B_0\varphi(0)\right)_m + \left(\hat{B}_0\psi(0), \tilde{B}_0\varphi(0)\right)_{n-m}.$$

Analogously

$$(\psi(T),\varphi(T))_n = \left(\bar{B}_1\psi(T), B_1\varphi(T)\right)_{n-m} + \left(\hat{B}_1\psi(T), \tilde{B}_1\varphi(T)\right)_m.$$

These two equalities yield representation (0.11); using the latter and (0.10), we obtain

$$\begin{aligned}
\int_0^T (L\varphi(t),\psi(t))_n \,\mathrm{d}t &= \left(\bar{B}_1\psi(T), B_1\varphi(T)\right)_{n-m} + \left(\hat{B}_1\psi(T), \tilde{B}_1\varphi(T)\right)_m \\
&- \left(\bar{B}_0\psi(0), B_0\varphi(0)\right)_m - \left(\hat{B}_0\psi(0), \tilde{B}_0\varphi(0)\right)_{n-m} + \int_0^T \left(\varphi(t), L^*\psi(t)\right)_n. \qquad (0.12)
\end{aligned}$$

In order to write the sum of the first four terms on the right-hand side of (0.12) in a scalar form, introduce the following notations:

$$\begin{pmatrix} U_{n+1}(\varphi) \\ \vdots \\ U_{2n-m}(\varphi) \end{pmatrix} := \tilde{B}_0\varphi(0) = \begin{pmatrix} \sum_{q=1}^{n} \tilde{b}_{1q}^{(0)}\varphi_q(0) \\ \vdots \\ \sum_{q=1}^{n} \tilde{b}_{n-m,q}^{(0)}\varphi_q(0) \end{pmatrix}, \tag{0.13}$$

$$\begin{pmatrix} U_{2n-m+1}(\varphi) \\ \vdots \\ U_{2n}(\varphi) \end{pmatrix} := \tilde{B}_1\varphi(T) = \begin{pmatrix} \sum_{q=1}^{n} \tilde{b}_{1q}^{(1)}\varphi_q(T) \\ \vdots \\ \sum_{q=1}^{n} \tilde{b}_{m,q}^{(1)}\varphi_q(T) \end{pmatrix}, \tag{0.14}$$

$$\begin{pmatrix} V_{2n}(\psi) \\ \vdots \\ V_{2n-m+1}(\psi) \end{pmatrix} := \bar{B}_0\psi(0) = \begin{pmatrix} \sum_{q=1}^{n} \bar{b}_{1q}^{(0)}\psi_q(0) \\ \vdots \\ \sum_{q=1}^{n} \bar{b}_{m,q}^{(0)}\psi_q(0) \end{pmatrix}, \tag{0.15}$$

$$\begin{pmatrix} V_{2n-m}(\psi) \\ \vdots \\ V_{n+1}(\psi) \end{pmatrix} := \bar{B}_1\psi(T) = \begin{pmatrix} \sum_{q=1}^{n} \bar{b}_{1q}^{(1)}\psi_q(T) \\ \vdots \\ \sum_{q=1}^{n} \bar{b}_{n-m,q}^{(1)}\psi_q(T) \end{pmatrix}, \tag{0.16}$$

$$\begin{pmatrix} V_{n}(\psi) \\ \vdots \\ V_{m+1}(\psi) \end{pmatrix} := \hat{B}_0\psi(0) = \begin{pmatrix} \sum_{q=1}^{n} \hat{b}_{1q}^{(0)}\psi_q(0) \\ \vdots \\ \sum_{q=1}^{n} \hat{b}_{n-m,q}^{(0)}\psi_q(0) \end{pmatrix}, \tag{0.17}$$

$$\begin{pmatrix} V_{m}(\psi) \\ \vdots \\ V_{1}(\psi) \end{pmatrix} := \hat{B}_1\psi(T) = \begin{pmatrix} \sum_{q=1}^{n} \hat{b}_{1q}^{(1)}\psi_q(T) \\ \vdots \\ \sum_{q=1}^{n} \hat{b}_{m,q}^{(1)}\psi_q(T) \end{pmatrix}. \tag{0.18}$$

Then the Equality (0.12) can be written as

$$\begin{aligned}
&\int_0^T \left(L\varphi(t),\psi(t)\right)_n \mathrm{d}t-\int_0^T \left(\varphi(t),L^*\psi(t)\right)_n \\
&=-U_1(\varphi)V_{2n}(\psi)-U_2(\varphi)V_{2n-1}(\psi)-\cdots-U_m(\varphi)V_{2n-m+1}(\psi) \\
&\quad+U_{m+1}(\varphi)V_{2n-m}(\psi)+U_{m+2}(\varphi)V_{2n-m-1}(\psi)+\cdots+U_n(\varphi)V_{n+1}(\psi) \\
&\quad-U_{n+1}(\varphi)V_n(\psi)-U_{n+2}(\varphi)V_{n-1}(\psi)-\cdots-U_{2n-m}(\varphi)V_{m+1}(\psi) \\
&\quad+U_{2n-m+1}(\varphi)V_m(\psi)+U_{2n-m+2}(\varphi)V_{m-1}(\psi)+\cdots+U_{2n}(\varphi)V_1(\psi).
\end{aligned} \tag{0.19}$$

Now we can introduce the notion of the adjoint BVP.

Definition 2.1 The homogeneous BVP

$$L^*\psi(t)=0, \quad t\in(0,T), \tag{0.20}$$

$$\hat{B}_0\psi(0)=0, \quad \hat{B}_1\psi(T)=0, \tag{0.21}$$

is called adjoint to homogeneous BVP (0.3), (0.4).

Definition 2.2 The inhomogeneous BVP

$$L^*\psi(t)=\tilde{f}(t), \quad t\in(0,T), \tag{0.22}$$

$$\hat{B}_0\psi(0)=\tilde{f}_0, \quad \hat{B}_1\psi(T)=\tilde{f}_1, \tag{0.23}$$

where $\tilde{f}\in\left(L^2(0,T)\right)^n$, $\tilde{f}_0\in\mathbb{R}^{n-m}$, $\tilde{f}_1\in\mathbb{R}^m$, is called adjoint to inhomogeneous BVP (0.1), (0.2).

The results contained in [13] [14] imply that the following statement is valid (for more detailed explanations see [9] , pp. 9-11).

Theorem 2.1 If homogeneous BVP (0.3), (0.4) has only the trivial solution, then the corresponding adjoint BVP (0.20), (0.21) also has only the trivial solution and inhomogeneous BVP (0.22), (0.23) has one and only one solution $\psi\in\left(H^1(0,T)\right)^n$ at any $\tilde{f}(t)\in\left(L^2(0,T)\right)^n$, $\tilde{f}_0\in\mathbb{R}^{n-m}$, $\tilde{f}_1\in\mathbb{R}^m$.

3. STATEMENT OF THE MINIMAX ESTIMATION PROBLEM AND ITS REDUCTION TO AN OPTIMAL CONTROL PROBLEM

Let a vector-function

$$y(t)=H(t)\varphi(t)+\xi(t), \tag{0.24}$$

with the values from the space $\mathbb{R}^l$ be observed on an interval $(\alpha,\beta)\subseteq(0,T)$; here $H(t)$ is an $l\times n$ matrix with the entries that are continuous functions on $[\alpha,\beta]$, and $\xi(t)\in\mathbb{R}^l$ is an unknown random vector process whose realizations enter observations (0.24).

Denote by V the set of random vector processes $\tilde{\xi}(t)$ with zero expectation $\mathbb{E}\tilde{\xi}(t)$ and second moments $\mathbb{E}\tilde{\xi}(t)^2$ integrable on (α,β) such that their correlation matrices $\tilde{R}(t,s)=\mathbb{E}\tilde{\xi}(t)\tilde{\xi}^{\mathrm{T}}(s)$ satisfy the inequality

$$\left\{\tilde{R}:\int_\alpha^\beta \mathrm{Sp}\left[Q(t)\tilde{R}(t,t)\right]\mathrm{d}t\le\varepsilon_1\right\}, \tag{0.25}$$

where $Q(t)$ is a positive definite matrix of dimension $l\times l$, the entries of $Q(t)$ and $Q^{-1}(t)$ are continuous on $[\alpha,\beta]$, ε_1 is a given positive number, $\mathrm{Sp}B:=\sum_{i=1}^l b_{ii}$ denotes the trace of the matrix $B=\left\{b_{ij}\right\}_{i,j=1}^l$.

Set

$$G=\Big\{\tilde{F}:=\left(\tilde{f}_0,\tilde{f}_1,\tilde{f}(\cdot)\right)\in\mathbb{R}^m\times\mathbb{R}^{n-m}\times\left(L^2(0,T)\right)^n:\left(Q_0\left(\tilde{f}_0-f_0^{(0)}\right),\tilde{f}_0-f_0^{(0)}\right)_m$$
$$+\left(Q_1\left(\tilde{f}_1-f_1^{(0)}\right),\tilde{f}_1-f_1^{(0)}\right)_{n-m}+\int_0^T\left(Q_2(t)\left(\tilde{f}(t)-f^{(0)}(t)\right),\tilde{f}(t)-f^{(0)}(t)\right)_n\mathrm{d}t\le\varepsilon_2\Big\}, \tag{0.26}$$

where $f_0^{(0)} \in \mathbb{R}^m$, $f_1^{(0)} \in \mathbb{R}^{n-m}$ are given vectors; $f^{(0)} \in \left(L^2(0,T)\right)^n$ is a given vector-function; Q_1 Q_0, and $Q_2(t)$ are positive definite matrices of dimensions $m \times m$, $(n-m)\times(n-m)$, and $n \times n$, respectively, the entries of $Q_2(t)$ and $Q_2^{-1}(t)$ are continuous on $[0,T]$, ε_2 is a given positive number.

Assume that the right-hand sides $f(\cdot)$, f_0, and f_1 of Equation (0.1) and boundary conditions (0.2) are not known exactly and it is known only that the element $F := (f_0, f_1, f(\cdot))$ belongs to a set G and, additionally, $\xi(t) \in V$. Further we also will assume, without loss of generality, that in (0.25) and (0.26) $\varepsilon_1 = \varepsilon_2 = 1$.

Let a vector-function $\varphi(t)$ be a solution to BVP (0.1), (0.2).

We will look for an estimation of the inner product

$$\left(a, \varphi(s)\right)_n \tag{0.27}$$

in the class of estimates linear with respect to observations that have the form

$$\begin{aligned}\widehat{\left(a, \varphi(s)\right)_n} &= \int_\alpha^\beta \left(u(t), y(t)\right)_l \mathrm{d}t + c \\ &= \int_\alpha^s \left(u_1(t), y(t)\right)_l \mathrm{d}t + \int_s^\beta \left(u_2(t), y(t)\right)_l \mathrm{d}t + c,\end{aligned} \tag{0.28}$$

where $s \in (\alpha, \beta)$ 1 and a is a vector belonging to $\mathbb{R}^n$, $u \in \left(L^2(\alpha,\beta)\right)^l$, $u_1 = u|_{(\alpha,s)}$, $u_2 = u|_{(s,T)}$, and c is certain constant. Then $u = (u_1, u_2) \in H := \left(L^2(\alpha,s)\right)^l \times \left(L^2(s,\beta)\right)^l = \left(L^2(\alpha,\beta)\right)^l$.

Definition 3.1 An estimate

$$\widehat{\widehat{(a,\varphi(s))_n}} = \int_\alpha^\beta (\hat{u}(t), y(t))_l \, dt + \hat{c}$$
$$= \int_\alpha^s (\hat{u}_1(t), y(t))_l \, dt + \int_s^\beta (\hat{u}_2(t), y(t))_l \, dt + \hat{c}$$

for which vector-function $\hat{u}(t) = (\hat{u}_1(t), \hat{u}_2(t))$ and constant $\hat{c}$ are determined from the condition

$$\inf_{u \in H, c \in \mathbf{R}} \sigma(u,c) = \sigma(\hat{u},\hat{c}),$$

where

$$\sigma(u,c) := \sup_{\tilde{F} \in G, \tilde{\xi} \in V} \mathbb{E}\left|(a,\tilde{\varphi}(s))_n - \widehat{(a,\tilde{\varphi}(s))_n}\right|^2,$$

$\tilde{\varphi}$ is a solution to BVP (0.1), (0.2) at $f(t) = \tilde{f}(t),\ f_0 = \tilde{f}_0,\ f_1 = \tilde{f}_1,$ and

$$\widehat{(a,\tilde{\varphi}(s))_n} = \int_\alpha^s (u_1(t), \tilde{y}(t))_l \, dt + \int_s^\beta (u_2(t), \tilde{y}(t))_l \, dt + c,$$

$$\tilde{y}(t) = H(t)\tilde{\varphi}(t) + \tilde{\xi}(t),$$

will be called a minimax mean square estimate of inner product $(a,\varphi(s))_n$. The quantity

$$\sigma = [\sigma(\hat{u},\hat{c})]^{1/2}$$

will be called an error of the minimax estimation.

We see that the minimax mean square estimate of inner product $(a,\varphi(s))_n$ is an estimate at which the maximum mean square estimation error calculated for the worst realization of perturbations attains its minimum.

We will show that solution to the minimax estimation problem is reduced to the solution of a certain optimal control problem.

For every fixed $u:=(u_1,u_2)\in H$ introduce vector-functions $z_1(\cdot;u)\in\left(H^1(0,\alpha)\right)^n$, $z_2(\cdot;u)\in\left(H^1(\alpha,s)\right)^n$, $z_3(\cdot;u)\in\left(H^1(s,\beta)\right)^n$, and $z_4(\cdot;u)\in\left(H^1(\beta,T)\right)^n$ as a solution to the following BVP:

$$\begin{aligned}
&L^* z_1(t;u)=0,\quad 0<t<\alpha,\quad \hat{B}_0 z_1(0;u)=0,\\
&L^* z_2(t;u)=-H^{\mathrm{T}}(t)u_1(t),\quad \alpha<t<s,\quad z_2(\alpha;u)=z_1(\alpha;u),\\
&L^* z_3(t;u)=-H^{\mathrm{T}}(t)u_2(t),\quad s<t<\beta,\quad z_3(s;u)=z_2(s;u)-a,\\
&L^* z_4(t;u)=0,\quad \beta<t<T,\quad z_4(\beta;u)=z_3(\beta;u),\quad \hat{B}_1 z_4(T;u)=0.
\end{aligned}\tag{0.29}$$

Lemma 3.1 Determination of the minimax mean square estimate of inner product $(a,\varphi(s))_n\cdot$ is equivalent to the problem of optimal control of the system described by BVP (0.29) with the cost function

$$\begin{aligned}
I(u)=&\left(Q_0^{-1}\bar{B}_0 z_1(0;u),\bar{B}_0 z_1(0;u)\right)_m+\left(Q_1^{-1}\bar{B}_1 z_4(T;u),\bar{B}_1 z_4(T;u)\right)_{n-m}\\
&+\int_0^\alpha\left(Q_2^{-1}(t)z_1(t;u),z_1(t;u)\right)_n dt+\int_\alpha^s\left(Q_2^{-1}(t)z_2(t;u),z_2(t;u)\right)_n dt\\
&+\int_s^\beta\left(Q_2^{-1}(t)z_3(t;u),z_3(t;u)\right)_n dt+\int_\beta^T\left(Q_2^{-1}(t)z_4(t;u),z_4(t;u)\right)_n dt\\
&+\int_\alpha^s\left(Q^{-1}(t)u_1(t),u_1(t)\right)_l dt+\int_s^\beta\left(Q^{-1}(t)u_2(t),u_2(t)\right)_l dt.
\end{aligned}\tag{0.30}$$

Proof. Show first that BVP (0.29) is uniquely solvable under the condition that functions $u_1(t)$ and $u_2(t)$ belong, respectively, to the spaces $\left(L^2(\alpha,s)\right)^l$ and $\left(L^2(s,\beta)\right)^l$.

Since homogeneous BVP (0.3), (0.4) has only the trivial solution, the BVP

$$L^{*}\psi(t)=g(t),\ 0<t<T,\ \hat{B}_{0}\psi(0)=0,\ \hat{B}_{1}\psi(T)=0 \tag{0.31}$$

has, in line with Theorem 2.1, the unique solution for any right-hand side, in particular, at

$$g(t)=g(t;u)=\begin{cases}0, & 0<t<\alpha;\\ -H^{\mathrm{T}}(t)u_{1}(t), & \alpha<t<s;\\ -H^{\mathrm{T}}(t)u_{2}(t), & s<t<\beta;\\ 0, & \beta<t<T.\end{cases} \tag{0.32}$$

Denote this solution by $\overline{z}(t;u)$ and its restrictions on intervals $(0,\alpha)$, (α,s), (s,β), and (β,T) by $\overline{z}_1(t;u)$, $\overline{z}_2(t;u)$, $\overline{z}_3(t;u)$, and $\overline{z}_4(t;u)$, respectively. Note that function $\overline{z}(t;u)$ is absolutely continuous on $[0,T]$ (see [15]).

Let us show that the problem

$$\begin{aligned}&L^{*}\overline{\overline{z}}^{(1)}(t)=0,\ 0<t<\alpha,\ \ \hat{B}_{0}\overline{\overline{z}}^{(1)}(0)=0,\\ &L^{*}\overline{\overline{z}}^{(2)}(t)=0,\ \alpha<t<s,\ \ \overline{\overline{z}}^{(2)}(\alpha;u)=\overline{\overline{z}}^{(1)}(\alpha;u),\\ &L^{*}\overline{\overline{z}}^{(3)}(t)=0,\ s<t<\beta,\ \ \overline{\overline{z}}^{(3)}(s;u)=\overline{\overline{z}}^{(2)}(s)-a,\\ &L^{*}\overline{\overline{z}}^{(4)}(t)=0,\ \beta<t<T,\ \ \overline{\overline{z}}^{(4)}(\beta)=\overline{\overline{z}}^{(3)}(\beta),\ \ \hat{B}_{1}\overline{\overline{z}}^{(4)}(T)=0\end{aligned} \tag{0.33}$$

has one and only one solution at any vector $a\in\mathbb{R}^{n}$.

Denote by $\overline{\overline{z}}_{i}^{(j)}(t),\ i=\overline{1,n},\ j=\overline{1,4},$ the coordinates of vector-function $\overline{\overline{z}}^{(j)}(t),\ j=\overline{1,4}.$ Let $y_{ik}(t),\ i,k=\overline{1,n}$ be the fundamental system of solutions of the equation system $L^{*}z(t)=0$ on $[0,T]$. Then we can represent functions $\overline{\overline{z}}_{i}^{(j)}(t),\ i=\overline{1,n},\ j=\overline{1,4},$ in the form

$$\overline{\overline{z}}_{i}^{(j)}(t)=\sum_{k=1}^{n}c_{k}^{(j)}y_{ik}(t),$$

where $c_k^{(j)}$ are constants. Taking into account the boundary conditions at the points $t = 0,T$ and transmission conditions at $t = \alpha, s, \beta$ in (0.33), we see that the solution to BVP (0.33) is equivalent to the solution of the following linear equation system with $4n$ unknowns $c_k^{(j)},\ k = \overline{1,n}$, $j = \overline{1,4}$:

$$\sum_{k=1}^{n} a_{ik}^{0} c_k^{(1)} = 0,\ \ i = \overline{1,n-m}, \tag{0.34}$$

$$\sum_{k=1}^{n} y_{ik}(\alpha)\left(c_k^{(1)} - c_k^{(2)}\right) = 0,\ \ i = \overline{1,n}, \tag{0.35}$$

$$\sum_{k=1}^{n} y_{ik}(s)\left(c_k^{(2)} - c_k^{(3)}\right) = a_i,\ \ i = \overline{1,n}, \tag{0.36}$$

$$\sum_{k=1}^{n} y_{ik}(\beta)\left(c_k^{(3)} - c_k^{(4)}\right) = 0,\ \ i = \overline{1,n}, \tag{0.37}$$

$$\sum_{k=1}^{n} a_{ik}^{1} c_k^{(4)} = 0,\ \ i = \overline{1,m}, \tag{0.38}$$

where

$$a_{ik}^{0} = \sum_{r=1}^{n} \hat{b}_{ir}^{(0)} y_{rk}(0),\ \ i = \overline{1,n-m},\ \ k = \overline{1,n},$$

$$a_{ik}^{1} = \sum_{s=1}^{n} \hat{b}_{is}^{(1)} y_{sk}(T),\ \ i = \overline{1,m},\ \ k = \overline{1,n},$$

$i=\overline{1,n}$, a_i, denote the coordinates of vector a, and $\hat{b}_{ir}^{(0)}$, $i=\overline{1,n-m}$, $r=\overline{1,n}$, and $\hat{b}_{is}^{(1)}$, $i=\overline{1,m}$, $s=\overline{1,n}$, denote the entries of matrices $\hat{B}_0$ and $\hat{B}_1$, respectively.

Show that system (0.34)-(0.38) is uniquely solvable at any vector $a\in\mathbb{R}^n$. To this end, note that homogeneous system (0.34)-(0.38) (at $a=0$) has only the trivial solution.

Indeed, setting $a=0$ in Equations (0.35) and (0.36), taking into account (0.37) and the fact that

$$\det\{y_{ik}(s)\}_{i,k=1}^{n}\neq 0\ \det\{y_{ik}(\alpha)\}_{i,k=1}^{n}\neq 0, \quad \text{and} \quad \det\{y_{ik}(\beta)\}_{i,k=1}^{n}\neq 0$$

because $y_{ik}(t)$, $i,k=\overline{1,n}$ is the fundamental system of solutions of the equation system $L^{*}z(t)=0$ on $[0,T]$, we obtain

$$c_k^{(1)}=c_k^{(2)}=c_k^{(3)}=c_k^{(4)}=:c_k.$$

Coefficients c_k satisfy Equations (0.34) and (0.38); therefore vector-function $\psi(t)$ with the components $\psi_i(t)=\sum_{k=1}^{n}c_k y_{ik}(t)$, $i=\overline{1,n}$ is a solution to conjugate BVP (0.20), (0.21) which has only the trivial solution $\psi(t)\equiv 0$ on $[0,T]$ by Theorem 2.1. This implies $c_k=0$, so the homogeneous linear equation system (0.34)- (0.38) (at $a=0$) has only the trivial solution. Consequently, system (0.34)-(0.38) and therefore BVP (0.33) which is equivalent to this system are uniquely solvable at any vector $a\in\mathbb{R}^n$. Then vector-functions $z_i(t;u)=\overline{z}_i(t;u)+\overline{\overline{z}}^{(i)}(t)$, $i=\overline{1,4}$, form the unique solution to BVP (0.29).

Show next that the determination of the minimax estimate of inner product $(a,\varphi(s))_n$ is equivalent to the problem of optimal control of the system described by BVP (0.29) with the cost function (0.30).

Using the second and third equations in (0.29) and the fact that $\tilde{\varphi}$ is a solution to BVP (0.1), (0.2) at $f(t)=\tilde{f}(t)$, $f_0=\tilde{f}_0$, and $f_1=\tilde{f}_1$, we easily obtain the relationships

$$-\int_\alpha^s \left(H^{\mathrm{T}}(t)u_1(t),\tilde{\varphi}(t)\right)_n dt=\left(z_2(\alpha;u),\tilde{\varphi}(\alpha)\right)_n-\left(z_2(s;u),\tilde{\varphi}(s)\right)_n+\int_\alpha^s\left(z_2(t;u),\tilde{f}(t)\right)_n dt,$$

$$-\int_s^\beta \left(H^{T}(t)u_2(t),\tilde{\varphi}(t)\right)_n dt=\left(z_3(s;u),\tilde{\varphi}(s)\right)_n-\left(z_3(\beta;u),\tilde{\varphi}(\beta)\right)_n+\int_s^\beta\left(z_3(t;u),\tilde{f}(t)\right)_n dt.$$

Taking into account the equalities

$$z_2(\alpha;u)=z_1(\alpha;u),\quad z_2(s;u)-z_3(s;u)=a,\quad z_3(\beta;u)=z_4(\beta;u),$$

$$\left(z_1(\alpha;u),\tilde{\varphi}(\alpha)\right)_n=\int_0^\alpha d\left(z_1(t;u),\tilde{\varphi}(t)\right)_n+\left(z_1(0;u),\tilde{\varphi}(0)\right)_n,$$

$$\left(z_4(\beta;u),\tilde{\varphi}(\beta)\right)_n=-\int_\beta^T d\left(z_4(t;u),\tilde{\varphi}(t)\right)_n+\left(z_4(T;u),\tilde{\varphi}(T)\right)_n,$$

and that

$$\left(z_1(0;u),\tilde{\varphi}(0)\right)_n=\left(\bar{B}_0z_1(0;u),B_0\tilde{\varphi}(0)\right)_m+\left(\hat{B}_0z_1(0;u),\tilde{B}_0\tilde{\varphi}(0)\right)_{n-m}=\left(\bar{B}_0z_1(0;u),\tilde{f}_0\right)_m,$$

$$\left(z_4(T;u),\tilde{\varphi}(T)\right)_n=\left(\bar{B}_1z_4(T;u),B_1\tilde{\varphi}(T)\right)_{n-m}+\left(\hat{B}_1z_4(T;u),\tilde{B}_1\tilde{\varphi}(T)\right)_m=\left(\bar{B}_1z_4(T;u),\tilde{f}_1\right)_{n-m},$$

(we refer to the reasoning on p. 4) we obtain

$$
\begin{aligned}
&\left(a,\tilde{\varphi}(s)\right)_n-\left(\widehat{a,\tilde{\varphi}(s)}\right)_n=\left(z_2(s;u),\tilde{\varphi}(s)\right)_n-\left(z_3(s;u),\tilde{\varphi}(s)\right)_n-\left(\widehat{a,\tilde{\varphi}(s)}\right)_n \\
&=\left(z_2(\alpha;u),\tilde{\varphi}(\alpha)\right)_n+\int_\alpha^s\left(H^{\mathrm{T}}(t)u_1(t),\tilde{\varphi}(t)\right)_n\mathrm{d}t+\int_\alpha^s\left(z_2(t;u),\tilde{f}(t)\right)_n\mathrm{d}t \\
&\quad-\left(z_3(\beta;u),\tilde{\varphi}(\beta)\right)_n+\int_s^\beta\left(H^{\mathrm{T}}(t)u_2(t),\tilde{\varphi}(t)\right)_n\mathrm{d}t+\int_s^\beta\left(z_3(t;u),\tilde{f}(t)\right)_n\mathrm{d}t \\
&\quad-\int_\alpha^s\left(u_1(t),\tilde{y}(t)\right)_l\mathrm{d}t-\int_s^\beta\left(u_2(t),\tilde{y}(t)\right)_l\mathrm{d}t-c \\
&=\int_0^\alpha\mathrm{d}\left(z_1(t;u),\tilde{\varphi}(t)\right)_n+\left(z_1(0;u),\tilde{\varphi}(0)\right)_n+\int_\alpha^s\left(H^{\mathrm{T}}(t)u_1(t),\tilde{\varphi}(t)\right)_n\mathrm{d}t \\
&\quad+\int_\alpha^s\left(z_2(t;u),\tilde{f}(t)\right)_n\mathrm{d}t+\int_\beta^T\mathrm{d}\left(z_4(t;u),\tilde{\varphi}(t)\right)_n-\left(z_4(T;u),\tilde{\varphi}(T)\right)_n \\
&\quad+\int_s^\beta\left(H^{\mathrm{T}}(t)u_2(t),\tilde{\varphi}(t)\right)_n\mathrm{d}t+\int_s^\beta\left(z_3(t;u),\tilde{f}(t)\right)_n\mathrm{d}t-\int_\alpha^s\left(H^{\mathrm{T}}(t)u_1(t),\tilde{\varphi}(t)\right)_n\mathrm{d}t \\
&\quad-\int_\alpha^s\left(u_1(t),\tilde{\xi}(t)\right)_l\mathrm{d}t-\int_s^\beta\left(H^{\mathrm{T}}(t)u_2(t),\tilde{\varphi}(t)\right)_n\mathrm{d}t-\int_s^\beta\left(u_2(t),\tilde{\xi}(t)\right)_l\mathrm{d}t-c \\
&=\int_0^\alpha\left(\frac{\mathrm{d}z_1(t;u)}{\mathrm{d}t},\tilde{\varphi}(t)\right)_n\mathrm{d}t+\int_0^\alpha\left(\frac{\mathrm{d}\tilde{\varphi}(t)}{\mathrm{d}t},z_1(t;u)\right)_n\mathrm{d}t+\left(\bar{B}_0z_1(0;u),\tilde{f}_0\right)_m \\
&\quad+\int_\alpha^s\left(z_2(t;u),\tilde{f}(t)\right)_n\mathrm{d}t+\int_\beta^T\left(\frac{\mathrm{d}z_4(t;u)}{\mathrm{d}t},\tilde{\varphi}(t)\right)_n\mathrm{d}t+\int_\beta^T\left(\frac{\mathrm{d}\tilde{\varphi}(t)}{\mathrm{d}t},z_4(t;u)\right)_n\mathrm{d}t \\
&\quad-\left(\bar{B}_1z_4(T;u),\tilde{f}_1\right)_{n-m}+\int_s^\beta\left(z_3(t;u),\tilde{f}(t)\right)_n\mathrm{d}t-\int_\alpha^s\left(u_1(t),\tilde{\xi}(t)\right)_l\mathrm{d}t-\int_s^\beta\left(u_2(t),\tilde{\xi}(t)\right)_l\mathrm{d}t-c.
\end{aligned}
$$

Taking into notice that

$$
\frac{\mathrm{d}z_1(t;u)}{\mathrm{d}t}=A^{\mathrm{T}}(t)z_1(t;u)\quad\text{on }(0,\alpha),\qquad\frac{\mathrm{d}z_4(t;u)}{\mathrm{d}t}=A^{\mathrm{T}}(t)z_4(t;u)\quad\text{on }(\beta,T)
$$

$$
\frac{\mathrm{d}\tilde{\varphi}(t)}{\mathrm{d}t}=\tilde{f}(t)-A(t)\tilde{\varphi}(t)\quad\text{on }(0,T),
$$

and therefore

$$
\begin{aligned}
&\int_0^\alpha\left(\frac{\mathrm{d}z_1(t;u)}{\mathrm{d}t},\tilde{\varphi}(t)\right)_n\mathrm{d}t+\int_0^\alpha\left(\frac{\mathrm{d}\tilde{\varphi}(t)}{\mathrm{d}t},z_1(t;u)\right)_n\mathrm{d}t \\
&=\int_0^\alpha\left(A^T(t)z_1(t;u),\tilde{\varphi}(t)\right)_n\mathrm{d}t+\int_0^\alpha\left(z_1(t;u),\tilde{f}(t)-A(t)\tilde{\varphi}(t)\right)_n\mathrm{d}t \\
&=\int_0^\alpha\left(z_1(t;u),\tilde{f}(t)\right)_n\mathrm{d}t,
\end{aligned}
$$

$$
\int_\beta^T\left(\frac{\mathrm{d}z_4(t;u)}{\mathrm{d}t},\tilde{\varphi}(t)\right)_n\mathrm{d}t+\int_\beta^T\left(\frac{\mathrm{d}\tilde{\varphi}(t)}{\mathrm{d}t},z_4(t;u)\right)_n\mathrm{d}t=\int_\beta^T\left(z_4(t;u),\tilde{f}(t)\right)_n\mathrm{d}t,
$$

we use the last equality to obtain

$$(a,\tilde{\varphi}(s))_n - \left(\widehat{a,\tilde{\varphi}(s)}\right)_n = \left(\bar{B}_0 z_1(0;u), \tilde{f}_0\right)_m + \int_0^T \left(z(t;u), \tilde{f}(t)\right)_n dt$$
$$-\left(\bar{B}_1 z_4(T;u), \tilde{f}_1\right)_{n-m} - \int_\alpha^s \left(u_1(t), \tilde{\xi}(t)\right)_l dt - \int_s^\beta \left(u_2(t), \tilde{\xi}(t)\right)_l dt - c =: \eta,$$
(0.39)

where

$$z(t;u) = \begin{cases} z_1(t;u), & 0 \le t \le \alpha; \\ z_2(t;u), & \alpha < t < s; \\ z_3(t;u), & s < t < \beta; \\ z_4(t;u), & \beta \le t \le T. \end{cases}$$

Recalling that $\tilde{\xi}(t)$ is a vector process with zero expectation, we use condition (0.25) and the known relationship $D\eta = \mathbb{E}\eta^2 - (\mathbb{E}\eta)^2$ that couples the dispersion $D\eta := \mathbb{E}[\eta - \mathbb{E}\eta]^2$ of random variable η with its expectation $\mathbb{E}\eta,$ to obtain

$$\mathbb{E}\eta = \left(\bar{B}_0 z_1(0;u), \tilde{f}_0\right)_m + \int_0^T \left(z(t;u), \tilde{f}(t)\right)_n dt - \left(\bar{B}_1 z_4(T;u), \tilde{f}_1\right)_{n-m} - c,$$

$$\eta - \mathbb{E}\eta = -\int_\alpha^s \left(u_1(t), \tilde{\xi}(t)\right)_l dt - \int_s^\beta \left(u_2(t), \tilde{\xi}(t)\right)_l dt,$$

$$D\eta = \mathbb{E}[\eta - \mathbb{E}\eta]^2 = \mathbb{E}\left[\int_\alpha^s \left(u_1(t), \tilde{\xi}(t)\right)_l dt + \int_s^\beta \left(u_2(t), \tilde{\xi}(t)\right)_l dt\right]^2,$$

$$\mathbb{E}\eta^2 = D\eta + (\mathbb{E}\eta)^2 = \mathbb{E}\left[\int_\alpha^s \left(u_1(t), \tilde{\xi}(t)\right)_l dt + \int_s^\beta \left(u_2(t), \tilde{\xi}(t)\right)_l dt\right]^2$$
$$+\left[\left(\bar{B}_0 z_1(0;u), \tilde{f}_0\right)_m + \int_0^T z\left((t;u), \tilde{f}(t)\right)_n dt - \left(\bar{B}_1 z_4(T;u), \tilde{f}_1\right)_{n-m} - c\right]^2,$$

which yields

$$\inf_{c\in\mathbf{R}^1}\sup_{\tilde{F}\in G,\tilde{\xi}\in V}\mathbb{E}\left[\left(a,\tilde{\varphi}(s)\right)_n-\left(\widehat{a,\tilde{\varphi}(s)}\right)_n\right]^2$$
$$=\inf_{c\in\mathbf{R}^1}\sup_{\tilde{F}\in G}\left[\left(\bar{B}_0 z_1(0;u),\tilde{f}_0\right)_m+\int_0^T\left(z(t;u),\tilde{f}(t)\right)_n\mathrm{d}t-\left(\bar{B}_1 z_4(T;u),\tilde{f}_1\right)_{n-m}-c\right]^2$$
$$+\sup_{\tilde{\xi}\in V}\mathbb{E}\left[\int_a^s\left(u_1(t),\tilde{\xi}(t)\right)_l\mathrm{d}t+\int_s^\beta\left(u_2(t),\tilde{\xi}(t)\right)_l\mathrm{d}t\right]^2. \tag{0.40}$$

In order to calculate the supremum on the right-hand side of (0.40) we apply the generalized CauchyBunyakovsky inequality [16] and write this inequality in the form convenient for further analysis: for any $f_0^{(1)},f_0^{(2)}\in R^m, f_1^{(1)},f_1^{(2)}\in R^{n-m}, f_1,f_2\in\left(L^2(0,T)\right)^n$, the generalized Cauchy-Bunyakovsky inequality holds

$$\left|\left(f_0^{(1)},f_0^{(2)}\right)_m+\left(f_1^{(1)},f_1^{(2)}\right)_{n-m}+\int_0^T\left(f_1(t),f_2(t)\right)_n\mathrm{d}t\right|$$
$$\le\left\{\left(Q_0^{-1}f_0^{(1)},f_0^{(1)}\right)_m+\left(Q_1^{-1}f_1^{(1)},f_1^{(1)}\right)_{n-m}+\int_0^T\left(Q_2^{-1}(t)f_1(t),f_1(t)\right)_n\mathrm{d}t\right\}^{\frac{1}{2}}$$
$$\times\left\{\left(Q_0f_0^{(2)},f_0^{(2)}\right)_m+\left(Q_1f_1^{(2)},f_1^{(2)}\right)_{n-m}+\int_0^T\left(Q_2(t)f_2(t),f_2(t)\right)_n\mathrm{d}t\right\}^{\frac{1}{2}},$$

in which the equality is attained at

$$f_0^{(2)}=\lambda Q_0^{-1}f_0^{(1)},\quad f_1^{(2)}=\lambda Q_1^{-1}f_1^{(1)},\quad f_2(t)=\lambda Q_2^{-1}f_1(t).$$

Setting in the generalized Cauchy-Bunyakovsky inequality

$$f_0^{(1)}=\bar{B}_0z_1(0;u),\quad f_1^{(1)}=-\bar{B}_1z_4(T;u),\quad f_1(t)=\tilde{z}(t;u),$$

$$f_0^{(2)}=\tilde{f}_0-f_0^{(0)},\quad f_1^{(2)}=\tilde{f}_1-f_1^{(0)},\quad f_2(t)=\tilde{f}(t)-f^{(0)}(t),$$

and denoting

$$Y:=\left(\bar{B}_0z_1(0;u),\tilde{f}_0-f_0^{(0)}\right)_m-\left(\bar{B}_1z_4(T;u),\tilde{f}_1-f_1^{(0)}\right)_{n-m}+\int_0^T\left(\tilde{z}(t;u),\tilde{f}(t)-f^{(0)}(t)\right)_n\mathrm{d}t$$

we obtain, in line with (2.7), the inequality

$$
\begin{aligned}
|Y| \le & \Big\{ \big(Q_0^{-1}\bar{B}_0 z_1(0;u), \bar{B}_0 z_1(0;u)\big)_m + \big(Q_1^{-1}\bar{B}_1 z_4(T;u), \bar{B}_1 z_4(T;u)\big)_{n-m} \\
& + \int_0^T \big(Q_2^{-1}(t) z(t;u), z(t;u)\big)_n \, dt \Big\}^{\frac{1}{2}} \times \Big\{ \big(Q_0(\tilde{f}_0 - f_0^{(0)}), \tilde{f}_0 - f_0^{(0)}\big)_m \\
& + \big(Q_1(\tilde{f}_1 - f_1^{(0)}), \tilde{f}_1 - f_1^{(0)}\big)_{n-m} + \int_0^T \big(Q_2(t)(\tilde{f}(t) - f^{(0)}(t)), \tilde{f}(t) - f^{(0)}(t)\big)_n \, dt \Big\}^{\frac{1}{2}} \\
\le & \Big\{ \big(Q_0^{-1}\bar{B}_0 z_1(0;u), \bar{B}_0 z_1(0;u)\big)_m + \big(Q_1^{-1}\bar{B}_1 z_4(T;u), \bar{B}_1 z_4(T;u)\big)_{n-m} \\
& + \int_0^T \big(Q_2^{-1}(t) \tilde{z}(t;u), \tilde{z}(t;u)\big)_n \, dt \Big\}^{\frac{1}{2}} := q,
\end{aligned}
$$

where the equality is attained at

$$
\tilde{f}_0 = \pm\frac{1}{q} Q_0^{-1}\bar{B}_0 z_1(0;u) + f_0^{(0)}, \quad \tilde{f}_1 = \mp\frac{1}{q} Q_1^{-1}\bar{B}_1 z_4(T;u) + f_1^{(0)}, \quad \tilde{f}(t) = \pm\frac{1}{q} Q_2(t) z(t;u) + f^{(0)}(t).
$$

Thus,

$$
\begin{aligned}
& \inf_{c\in\mathbf{R}^1} \sup_{\tilde{F}\in G} \Big[\big(\bar{B}_0 z_1(0;u), \tilde{f}_0\big)_m + \int_0^T \big(z(t;u), \tilde{f}(t)\big)_n \, dt - \big(\bar{B}_1 z_4(T;u), \tilde{f}_1\big)_{n-m} - c \Big]^2 \\
& = \inf_{c\in R^1} \sup_{\tilde{F}\in G_0} \Big[\big(\bar{B}_0 z_1(0;u), \tilde{f}_0 - f_0^{(0)}\big)_m - \big(\bar{B}_1 z_4(T;u), \tilde{f}_1 - f_1^{(0)}\big)_{n-m} + \int_0^T \big(z(t;u), \tilde{f}(t) - f^{(0)}(t)\big)_n \, dt \\
& \quad + \big(\bar{B}_0 z_1(0;u), f_0^{(0)}\big)_m - \big(\bar{B}_1 z_4(T;u), f_1^{(0)}\big)_{n-m} + \int_0^T \big(\tilde{z}(t;u), f^{(0)}(t)\big)_n \, dt - c \Big]^2 \\
& = \inf_{c\in R^1} \sup_{|Y|\le q} \Big[Y + \big(\bar{B}_0 z_1(0;u), f_0^{(0)}\big)_m - \big(\bar{B}_1 z_4(T;u), f_1^{(0)}\big)_{n-m} + \int_0^T \big(z(t;u), f^{(0)}(t)\big)_n \, dt - c \Big]^2 = q^2 \\
& = \big(Q_0^{-1}\bar{B}_0 z_1(0;u), \bar{B}_0 z_1(0;u)\big)_m + \big(Q_1^{-1}\bar{B}_1 z_4(T;u), \bar{B}_1 z_4(T;u)\big)_{n-m} + \int_0^T \big(Q_2^{-1}(t) z(t;u), z(t;u)\big)_n \, dt
\end{aligned}
\tag{0.41}
$$

at $c = \big(\bar{B}_0 z_1(0;u), f_0^{(0)}\big)_m - \big(\bar{B}_1 z_4(T;u), f_1^{(0)}\big)_{n-m} + \int_0^T \big(z(t;u), f^{(0)}(t)\big)_n \, dt.$

Calculate the second term on the right-hand side of (0.40). Applying the generalized Cauchy-Bunyakovsky inequality, we have

$$
\begin{aligned}
& \mathbb{E}\Big[\int_\alpha^s \big(u_1(t), \tilde{\xi}(t)\big)_l \, dt + \int_s^\beta \big(u_2(t), \tilde{\xi}(t)\big)_l \, dt \Big]^2 = \mathbb{E}\Big[\int_\alpha^\beta \big(u(t), \tilde{\xi}(t)\big)_l \, dt \Big]^2 \\
& \le \mathbb{E}\Big[\int_\alpha^\beta \big(Q^{-1}(t) u(t), u(t)\big)_l \, dt \cdot \int_\alpha^\beta \big(Q(t)\xi(t), \xi(t)\big)_l \, dt \Big] \\
& = \int_\alpha^\beta \big(Q^{-1}(t) u(t), u(t)\big)_l \, dt \cdot \int_\alpha^\beta \mathbb{E}\big(Q(t)\tilde{\xi}(t), \tilde{\xi}(t)\big)_l \, dt.
\end{aligned}
\tag{0.42}
$$

Here $\mathbb{E}$ can be placed under the integral sign according to the Fubini theorem because we assume that $\tilde{\xi}(t)$ is a random process of the integrable second moment. Transform the last factor on the right-hand side of (0.42):

$$\begin{aligned}&\int_\alpha^\beta \mathbb{E}\left(Q(t)\tilde{\xi}(t),\tilde{\xi}(t)\right)_l \mathrm{d}t\\ &=\int_\alpha^\beta \mathbb{E}\left(\sum_{i=1}^{l}\left(Q(t)\tilde{\xi}(t)\right)_i \tilde{\xi}_i(t)\right)\mathrm{d}t=\int_\alpha^\beta \mathbb{E}\left(\sum_{i=1}^{l}\sum_{k=1}^{l} q_{ik}\tilde{\xi}_k(t)\tilde{\xi}_i(t)\right)\mathrm{d}t\\ &=\int_\alpha^\beta \sum_{i=1}^{l}\sum_{k=1}^{l} q_{ik}\mathbb{E}\left(\tilde{\xi}_k(t)\tilde{\xi}_i(t)\right)\mathrm{d}t=\int_\alpha^\beta Sp\left[Q(t)\tilde{R}(t,t)\right]\mathrm{d}t.\end{aligned}$$

Taking into account that (0.25) holds, we see that (0.42) yields

$$\begin{aligned}&\sup_{\tilde{\xi}\in V}\mathbb{E}\left[\int_\alpha^s \left(u_1(t),\tilde{\xi}(t)\right)_l \mathrm{d}t+\int_s^\beta \left(u_2(t),\tilde{\xi}(t)\right)_l \mathrm{d}t\right]^2\\ &\le \int_\alpha^s \left(Q^{-1}(t)u_1(t),u_1(t)\right)_l \mathrm{d}t+\int_s^\beta \left(Q^{-1}(t)u_2(t),u_2(t)\right)_l \mathrm{d}t.\end{aligned} \tag{0.43}$$

It is not difficult to check that here, the equality sign is attained at the element

$$\tilde{\xi}(t)=\begin{cases}\dfrac{\eta Q^{-1}(t)u_1(t)}{\left[\int_\alpha^s \left(Q^{-1}(t)u_1(t),u_1(t)\right)_l \mathrm{d}t+\int_s^\beta \left(Q^{-1}(t)u_2(t),u_2(t)\right)_l \mathrm{d}t\right]^{1/2}}, & \alpha<t<s;\\[2ex] \dfrac{\eta Q^{-1}(t)u_2(t)}{\left[\int_\alpha^s \left(Q^{-1}(t)u_1(t),u_1(t)\right)_l \mathrm{d}t+\int_s^\beta \left(Q^{-1}(t)u_2(t),u_2(t)\right)_l \mathrm{d}t\right]^{1/2}}, & s<t<\beta,\end{cases}$$

where η is a random variable such that $\mathbb{E}\eta=0$ and $\mathbb{E}\eta^2=1.$ We conclude that statement of the lemma follows now from (0.40), (0.41), and (0.43).

4. REPRESENTATIONS FOR MINIMAX MEAN SQUARE ESTIMATES OF FUNCTIONALS FROM SOLUTIONS TO TWO-POINT BOUNDARY VALUE PROBLEMS AND ESTIMATION ERRORS

In this section we prove the theorem concerning general form of minimax mean square estimates. Solving optimal control problem (0.29), (0.30), we arrive at the following result.

Theorem 4.1 The minimax mean square estimate of expression $(a,\varphi(s))$ has the form

$$\widehat{\widehat{(a,\varphi(s))_n}} = \int_\alpha^s (\hat{u}_1(t), y(t))_l \, dt + \int_s^\beta (\hat{u}_2(t), y(t))_l \, dt$$

where

$$\hat{u}_1(t) = Q(t)H(t)p_2(t), \quad \hat{u}_2(t) = Q(t)H(t)p_3(t), \quad (0.44)$$

$$\hat{c} = \left(\overline{B}_0 z_1(0), f_0^{(0)}\right)_m - \left(\overline{B}_1 z_4(T), f_1^{(0)}\right)_{n-m} + \int_0^T \left(z(t), f^{(0)}(t)\right)_n dt,$$

$$z(t) = \begin{cases} z_1(t), & 0 \le t \le \alpha; \\ z_2(t), & \alpha < t < s; \\ z_3(t), & s < t < \beta; \\ z_4(t), & \beta \le t \le T, \end{cases} \quad (0.45)$$

and vector-functions $p_i(t)$ and $z_i(t)$, $i = \overline{1,4}$, are determined from the solution to the problem

$$\begin{aligned}
&L^{*}z_1 t=0,\quad 0<t<\alpha,\quad \hat{B}_0 z_1(0)=0,\\
&L^{*}z_2(t)=-H^{T}(t)Q(t)H(t)p_2(t),\quad \alpha<t<s,\quad z_2(\alpha)=z_1(\alpha),\\
&L^{*}z_3(t)=-H^{T}(t)Q(t)H(t)p_3(t),\quad s<t<\beta,\quad z_3(s)=z_2(s)-a,\\
&L^{*}z_4(t)=0,\quad \beta<t<T,\quad z_4(\beta)=z_3(\beta),\quad \hat{B}_1 z_4(T)=0,\\
&Lp_1(t)=Q_2^{-1}(t)z_1(t),\quad 0<t<\alpha,\quad B_0 p_1(0)=Q_0^{-1}\overline{B}_0 z_1(0),\\
&Lp_2(t)=Q_2^{-1}(t)z_2(t),\quad \alpha<t<s,\quad p_2(\alpha)=p_1(\alpha),\\
&Lp_3(t)=Q_2^{-1}(t)z_3(t),\quad s<t<\beta,\quad p_3(s)=p_2(s),\\
&Lp_4(t)=Q_2^{-1}(t)z_4(t),\quad \beta<t<T,\quad p_4(\beta)=p_3(\beta),\\
&B_1 p_4(T)=-Q_1^{-1}\overline{B}_1 z_4(T).
\end{aligned} \tag{0.46}$$

Here $z_1,p_1\in\left(H^1(0,\alpha)\right)^n$, $z_2,p_2\in\left(H^1(\alpha,s)\right)^n$, $z_3,p_3\in\left(H^1(s,\beta)\right)^n$, and $z_4,p_4\in\left(H^1(\beta,T)\right)^n$. The minimax estimation error

$$\sigma=\left(a,p_2(s)\right)_n^{1/2}. \tag{0.47}$$

Problem (0.46) is uniquely solvable.

Proof. We will solve optimal control problem (0.29), (0.30). Represent solutions $z_i(x;u)$, $i=\overline{1,4}$, of problem (0.29) as $z_i(t;u)=\overline{z}_i(t;u)+\overline{\overline{z}}_i(t)$, where $\overline{z}_1(t;u)$, $\overline{z}_2(t;u)$, $\overline{z}_3(t;u)$, $\overline{z}_4(t;u)$ and $\overline{\overline{z}}_1(t)$, $\overline{\overline{z}}_2(t)$, $\overline{\overline{z}}_3(t)$, $\overline{\overline{z}}_4(t)$ denote the solutions to this problem at $a=0$ and $u_1\equiv 0,\ u_2\equiv 0$, respectively. Then function (0.30) can be represented in the form

$$I(u)=\tilde{I}(u)+2L(u)+A,$$

where

$$\begin{aligned}\tilde{I}(u)=&\left(Q_0^{-1}\bar{B}_0\bar{z}_1(0;u),\bar{B}_0\bar{z}_1(0;u)\right)_m+\left(Q_1^{-1}\bar{B}_1\bar{z}_4(T;u),\bar{B}_1\bar{z}_4(T;u)\right)_{n-m}\\&+\int_0^\alpha\left(Q_2^{-1}(t)\bar{z}_1(t;u),\bar{z}_1(t;u)\right)_n dt+\int_\alpha^s\left(Q_2^{-1}(t)\bar{z}_2(t;u),\bar{z}_2(t;u)\right)_n dt\\&+\int_s^\beta\left(Q_2^{-1}(t)\bar{z}_3(t;u),\bar{z}_3(t;u)\right)_n dt+\int_\beta^T\left(Q_2^{-1}(t)\bar{z}_4(t;u),\bar{z}_4(t;u)\right)_n dt\\&+\int_\alpha^s\left(Q^{-1}(t)u_1(t),u_1(t)\right)_l dt+\int_s^\beta\left(Q^{-1}(t)u_2(t),u_2(t)\right)_l dt,\end{aligned}$$

$$\begin{aligned}L(u)=&\left(Q_0^{-1}\bar{B}_0\bar{z}_1(0;u),\bar{B}_0\bar{\bar{z}}_1(0)\right)_m+\left(Q_1^{-1}\bar{B}_1\bar{z}_4(T;u),\bar{B}_1\bar{\bar{z}}_4(T)\right)_{n-m}\\&+\int_0^\alpha\left(Q_2^{-1}(t)\bar{z}_1(t;u),\bar{\bar{z}}_1(t)\right)_n dt+\int_\alpha^s\left(Q_2^{-1}(t)\bar{z}_2(t;u),\bar{\bar{z}}_2(t)\right)_n dt\\&+\int_s^\beta\left(Q_2^{-1}(t)\bar{z}_3(t;u),\bar{\bar{z}}_3(t)\right)_n dt+\int_\beta^T\left(Q_2^{-1}(t)\bar{z}_4(t;u),\bar{\bar{z}}_4(t)\right)_n dt,\end{aligned}$$

$$\begin{aligned}A=&\left(Q_0^{-1}\bar{B}_0\bar{\bar{z}}_1(0),\bar{B}_0\bar{\bar{z}}_1(0)\right)_m+\left(Q_1^{-1}\bar{B}_1\bar{\bar{z}}_4(T),\bar{B}_1\bar{\bar{z}}_4(T)\right)_{n-m}\\&+\int_0^\alpha\left(Q_2^{-1}(t)\bar{\bar{z}}_1(t),\bar{\bar{z}}_1(t)\right)_n dt+\int_\alpha^s\left(Q_2^{-1}(t)\bar{\bar{z}}_2(t),\bar{\bar{z}}_2(t)\right)_n dt\\&+\int_s^\beta\left(Q_2^{-1}(t)\bar{\bar{z}}_3(t),\bar{\bar{z}}_3(t)\right)_n dt+\int_\beta^T\left(Q_2^{-1}(t)\bar{\bar{z}}_4(t),\bar{\bar{z}}_4(t)\right)_n dt.\end{aligned}$$

Since solution $\bar{z}(t;u)$ of BVP (0.31) is continuous[2] with respect to right-hand side $g(t;u)$ defined by

(0.32), the function $u\to\bar{z}(\cdot;u)$ is a linear bounded operator mapping the space $H=\left(L^2(\alpha,s)\right)^l\times\left(L^2(s,\beta)\right)^l$ to

$$H_1:=\left(H^1(0,\alpha)\right)^n\times\left(H^1(\alpha,s)\right)^n\times\left(H^1(s,\beta)\right)^n\times\left(H^1(\beta,T)\right)^n.$$

Thus, $\tilde{I}(u)$ is a continuous quadratic form corresponding to a symmetric continuous bilinear form

$$\pi(u,v):=\left(Q_0^{-1}\overline{B}_0\overline{z}_1(0;u),\overline{B}_0\overline{z}_1(0;v)\right)_m+\left(Q_1^{-1}\overline{B}_1\overline{z}_4(T;u),\overline{B}_1\overline{z}_4(T;v)\right)_{n-m}$$
$$+\int_0^\alpha\left(Q_2^{-1}(t)\overline{z}_1(t;u),\overline{z}_1(t;v)\right)_n\mathrm{d}t+\int_\alpha^s\left(Q_2^{-1}(t)\overline{z}_2(t;u),\overline{z}_2(t;v)\right)_n\mathrm{d}t$$
$$+\int_s^\beta\left(Q_2^{-1}(t)\overline{z}_3(t;u),\overline{z}_3(t;v)\right)_n\mathrm{d}t+\int_\beta^T\left(Q_2^{-1}(t)\overline{z}_4(t;u),\overline{z}_4(t;v)\right)_n\mathrm{d}t$$
$$+\int_\alpha^s\left(Q^{-1}(t)u_1(t),v_1(t)\right)_l\mathrm{d}t+\int_s^\beta\left(Q^{-1}(t)u_2(t),v_2(t)\right)_l\mathrm{d}t,$$

$L(u)$ is a linear continuous functional defined on H, and A is a constant independent of u. We have

$$\tilde{I}(u)=\tilde{I}(u_1,u_2)\geq\int_\alpha^s\left(Q^{-1}(t)u_1(t),u_1(t)\right)_l\mathrm{d}t+\int_s^\beta\left(Q^{-1}(t)u_2(t),u_2(t)\right)_l\mathrm{d}t$$
$$\geq c\|u\|_H^2,\quad c=\text{const.}$$

Using Theorem 1.1 from [17] , we conclude that there is one and only one element $\hat{u}=(\hat{u}_1,\hat{u}_2)\in H$ such that

$$I(\hat{u})=I(\hat{u}_1,\hat{u}_2)=\inf_{(u_1,u_2)\in H}I(u_1,u_2).$$

Therefore

$$\frac{\mathrm{d}}{\mathrm{d}\tau}I\left(\hat{u}_1+\tau v_1,\hat{u}_2+\tau v_2\right)\Big|_{\tau=0}=0\quad\forall v=(v_1,v_2)\in H.$$

Taking into consideration the latter equality, (0.30), and designations on p. 11, we obtain

$$0=\frac{1}{2}\frac{\mathrm{d}I(\hat{u}+\tau v)}{\mathrm{d}\tau}\Big|_{\tau=0}=\int_\alpha^s\left(Q^{-1}(t)\hat{u}_1,v_1\right)_l\mathrm{d}t+\int_s^\beta\left(Q^{-1}(t)\hat{u}_2,v_2\right)_l\mathrm{d}t$$
$$+\left(Q_0^{-1}\overline{B}_0z_1(0;\hat{u}),\overline{B}_0\overline{z}_1(0;v)\right)_m+\left(Q_1^{-1}\overline{B}_1z_4(T;\hat{u}),\overline{B}_1\overline{z}_4(T;v)\right)_{n-m}$$
$$+\int_0^\alpha\left(Q_2^{-1}(t)z_1(t;\hat{u}),\overline{z}_1(t;v)\right)_n\mathrm{d}t+\int_\alpha^s\left(Q_2^{-1}(t)z_2(t;\hat{u}),\overline{z}_2(t;v)\right)_n\mathrm{d}t$$
$$+\int_s^\beta\left(Q_2^{-1}(t)z_3(t;\hat{u}),\overline{z}_3(t;v)\right)_n\mathrm{d}t+\int_\beta^T\left(Q_2^{-1}(t)z_4(t;\hat{u}),\overline{z}_4(t;v)\right)_n\mathrm{d}t.$$

(0.48)

Introduce functions $p_1 \in \left(H^1(0,\alpha)\right)^n$, $p_2 \in \left(H^1(\alpha,s)\right)^n$, $p_3 \in \left(H^1(s,\beta)\right)^n$, and $p_4 \in \left(H^1(\beta,T)\right)^n$ as a unique solution to the following problem[3]:

$$
\begin{aligned}
&Lp_1(t) = Q_2^{-1}(t) z_1(t;\hat{u}), \quad 0<t<\alpha, \quad B_0 p_1(0) = Q_0^{-1}\overline{B}_0 z_1(0;\hat{u}),\\
&Lp_2(t) = Q_2^{-1}(t) z_2(t;\hat{u}), \quad \alpha<t<s, \quad p_2(\alpha) = p_1(\alpha),\\
&Lp_3(t) = Q_2^{-1}(t) z_3(t;\hat{u}), \quad s<t<\beta, \quad p_3(s) = p_2(s),\\
&Lp_4(t) = Q_2^{-1}(t) z_4(t;\hat{u}), \quad \beta<t<T,\\
&p_4(\beta) = p_3(\beta), \quad B_1 p_4(T) = -Q_1^{-1}\overline{B}_1 z_4(T;\hat{u}).
\end{aligned}
\tag{0.49}
$$

Now transform the sum of the last for terms on the right-hand side of (0.48) taking into notice that $\left(\bar{z}_1(0;v), p_1(0)\right)_n = \left(\overline{B}_0\bar{z}_1(0;v), B_0 p_1(0)\right)_m$ and $\left(\bar{z}_4(T;v), p_4(T)\right)_n = \left(\overline{B}_1\bar{z}_4(T;v), B_1 p_4(T)\right)_{n-m}$. We have

$$
\begin{aligned}
&\int_0^\alpha \left(Q_2^{-1}(t) z_1(t;\hat{u}), \bar{z}_1(t;v)\right)_n dt + \int_\alpha^s \left(Q_2^{-1}(t) z_2(t;\hat{u}), \bar{z}_2(t;v)\right)_n dt\\
&+\int_s^\beta \left(Q_2^{-1}(t) z_3(t;\hat{u}), \bar{z}_3(t;v)\right)_n dt + \int_\beta^T \left(Q_2^{-1}(t) z_4(t;\hat{u}), \bar{z}_4(t;v)\right)_n dt\\
&= \int_0^\alpha \left(Lp_1(t), \bar{z}_1(t;v)\right)_n dt + \int_\alpha^s \left(Lp_2(t), \bar{z}_2(t;v)\right)_n dt\\
&\quad + \int_s^\beta \left(Lp_3(t), \bar{z}_3(t;v)\right)_n dt + \int_\beta^T \left(Lp_4(t), \bar{z}_4(t;v)\right)_n dt\\
&= \int_0^\alpha \left(p_1(t), L^*\bar{z}_1(t;v)\right)_n dt + \left(\bar{z}_1(\alpha;v), p_1(\alpha)\right)_n - \left(\bar{z}_1(0;v), p_1(0)\right)_n\\
&\quad + \int_\alpha^s \left(p_2(t), L^*\bar{z}_2(t;v)\right)_n dt + \left(\bar{z}_2(s;v), p_2(s)\right)_n - \left(\bar{z}_2(\alpha;v), p_2(\alpha)\right)_n\\
&\quad + \int_s^\beta \left(p_3(t), L^*\bar{z}_3(t;v)\right)_n dt + \left(\bar{z}_3(\beta;v), p_3(\beta)\right)_n - \left(\bar{z}_3(s;v), p_3(s)\right)_n\\
&\quad + \int_\beta^T \left(p_4(t), L^*\bar{z}_4(t;v)\right)_n dt + \left(\bar{z}_4(T;v), p_4(T)\right)_n - \left(\bar{z}_4(\beta;v), p_4(\beta)\right)_n\\
&= -\left(\bar{z}_1(0;v), p_1(0)\right)_n + \left(\bar{z}_4(T;v), p_4(T)\right)_n - \int_\alpha^s \left(p_2(t), H^T(t) v_1(t)\right)_n dt\\
&\quad - \int_s^\beta \left(p_3(t), H^T(t) v_2(t)\right)_n dt + \left(\bar{z}_2(s;v) - \bar{z}_3(s;v), p_2(s)\right)_n
\end{aligned}
$$

$$= -\left(\bar{B}_0\bar{z}_1(0;v), B_0 p_1(0)\right)_m + \left(\bar{B}_1\bar{z}_4(T;v), B_1 p_4(T)\right)_{n-m}$$
$$-\int_\alpha^s \left(H(t)p_2(t), v_1(t)\right)_l \mathrm{d}t - \int_s^\beta \left(H(t)p_3(t), v_2(t)\right)_l \mathrm{d}t$$
$$= -\left(\bar{B}_0\bar{z}_1(0;v), Q_0^{-1}\bar{B}_0 z_1(0;\hat{u})\right)_m - \left(B_1\bar{z}_4(T;v), Q_1^{-1}\bar{B}_1 z_4(T;\hat{u})\right)_{n-m}$$
$$-\int_\alpha^s \left(H(t)p_2(t), v_1(t)\right)_l \mathrm{d}t - \int_s^\beta \left(H(t)p_3(t), v_2(t)\right)_l \mathrm{d}t$$

(0.50)

From Equalities (0.48)-(0.50) it follows that

$$\int_\alpha^s \left(Q^{-1}(t)\hat{u}_1(t), v_1(t)\right)_l \mathrm{d}t + \int_s^\beta \left(Q^{-1}(t)\hat{u}_2(t), v_2(t)\right)_l \mathrm{d}t$$
$$= \int_\alpha^s \left(H(t)p_2(t), v_1(t)\right)_l \mathrm{d}t + \int_s^\beta \left(H(t)p_3(t), v_2(t)\right)_l \mathrm{d}t,$$

so that

$$Q^{-1}(t)\hat{u}_1(t) = H(t)p_2(t),\quad Q^{-1}(t)\hat{u}_2(t) = H(t)p_3(t),$$
$$\hat{u}_1(t) = Q(t)H(t)p_2(t),\quad \hat{u}_2(t) = Q(t)H(t)p_3(t). \tag{0.51}$$

Functions $p_1(t)$, $p_2(t)$, $p_3(t)$, and $p_4(t)$ are absolutely continuous on segments $[0,\alpha]$, $[\alpha,s]$,

$[s,\beta]$, and $[\beta,T]$, respectively, as solutions to BVP (0.49); therefore, functions $\hat{u}_1(t)$ and $\hat{u}_2(t)$ that perform optimal control are continuous on $[\alpha,s]$ and $[s,\beta]$. Replacing in (0.29) functions $u_1(t)$ and $u_2(t)$ by $\hat{u}_1(t)$ and $\hat{u}_2(t)$ defined by formulas (0.51) and denoting $z_i(t) = z_i(t;\hat{u})$, $i = \overline{1,4}$, we arrive at problem (0.46) and equalities (0.44).

Taking into consideration the way this problem was formulated we can state that its unique solvability follows from the fact that functional (0.30) has one minimum point $\hat{u}$.

Now let us prove representation (0.47). Substituting into formula $\sigma^2 = I(\hat{u})$ expressions (0.44) for $\hat{u}_1(t)$

and $\hat{u}_2(t)$, we have

$$\begin{aligned}
\sigma^2 &= \left(Q_0^{-1}\bar{B}_0 z_1(0), \bar{B}_0 z_1(0)\right)_m + \left(Q_1^{-1}\bar{B}_1 z_4(T), \bar{B}_1 z_4(T)\right)_{n-m} \\
&+ \int_0^\alpha \left(Q_2^{-1}(t) z_1(t), z_1(t)\right)_n \mathrm{d}t + \int_\alpha^s \left(Q_2^{-1}(t) z_2(t), z_2(t)\right)_n \mathrm{d}t \\
&+ \int_s^\beta \left(Q_2^{-1}(t) z_3(t), z_3(t)\right)_n \mathrm{d}t + \int_\beta^T \left(Q_2^{-1}(t) z_4(t), z_4(t)\right)_n \mathrm{d}t \\
&+ \int_\alpha^s \left(Q(t) H(t) p_2(t), H(t) p_2(t)\right)_l \mathrm{d}t \\
&+ \int_s^\beta \left(Q(t) H(t) p_3(t), H(t) p_3(t)\right)_l \mathrm{d}t.
\end{aligned} \tag{0.52}$$

Next, we can apply the reasoning similar to that on p. 4 and use (0.46) to obtain

$$\left(z_1(0), p_1(0)\right)_n = \left(\bar{B}_0 z_1(0), B_0 p_1(0)\right)_m = \left(\bar{B}_0 z_1(0), Q_0^{-1}\bar{B}_0 z_1(0)\right)_m,$$

which yields

$$\begin{aligned}
&\int_0^\alpha \left(Q_2^{-1}(t) z_1(t), z_1(t)\right)_n \mathrm{d}t + \int_\alpha^s \left(Q(t) H(t) p_2(t), H(t) p_2(t)\right)_l \mathrm{d}t \\
&= \int_0^\alpha \left(L p_1(t), z_1(t)\right)_n \mathrm{d}t - \int_\alpha^s \left(L^* z_2(t), p_2(t)\right)_n \mathrm{d}t \\
&= \int_0^\alpha \left(p_1(t), L^* z_1(t)\right)_n \mathrm{d}t + \left(z_1(\alpha), p_1(\alpha)\right)_n - \left(z_1(0), p_1(0)\right)_n \\
&\quad - \int_\alpha^s \left(z_2(t), L p_2(t)\right)_n \mathrm{d}t - \left(z_2(\alpha), p_2(\alpha)\right)_n + \left(z_2(s), p_2(s)\right)_n \\
&= -\left(z_1(0), p_1(0)\right)_n - \int_\alpha^s \left(z_2(t), Q_2^{-1}(t) z_2(t)\right)_n \mathrm{d}t + \left(z_2(s), p_2(s)\right)_n \\
&= -\int_\alpha^s \left(Q_2^{-1}(t) z_2(t), z_2(t)\right)_n \mathrm{d}t - \left(\bar{B}_0 z_1(0), Q_0^{-1}\bar{B}_0 z_1(0)\right)_m + \left(z_2(s), p_2(s)\right)_n.
\end{aligned} \tag{0.53}$$

In a similar manner, using the equality

$$\left(z_4(T),p_4(T)\right)_n=\left(\overline{B}_1z_4(T),B_1p_4(T)\right)_{n-m}=-\left(\overline{B}_1z_4(T),Q_1^{-1}\overline{B}_1z_4(T)\right)_{n-m},$$

we obtain

$$\begin{aligned}
&\int_\beta^T\left(Q_2^{-1}(t)z_4(t),z_4(t)\right)_n\mathrm{d}t+\int_s^\beta\left(Q(t)H(t)p_3(t),H(t)p_3(t)\right)_l\mathrm{d}t\\
&=\int_\beta^T\left(Lp_4(t),z_4(t)\right)_n\mathrm{d}t-\int_s^\beta\left(L^*z_3(t),p_3(t)\right)_n\mathrm{d}t\\
&=\int_\beta^T\left(p_4(t),L^*z_4(t)\right)_n\mathrm{d}t+\left(p_4(T),z_4(T)\right)_n-\left(p_4(\beta),z_4(\beta)\right)_n\\
&\quad-\int_s^\beta\left(z_3(t),Lp_3(t)\right)_n\mathrm{d}t+\left(z_3(\beta),p_3(\beta)\right)_n-\left(z_3(s),p_3(s)\right)_n\\
&=\left(p_4(T),z_4(T)\right)_n-\int_s^\beta\left(z_3(t),Q_2^{-1}(t)z_3(t)\right)_n\mathrm{d}t-\left(z_3(s),p_2(s)\right)_n\\
&=-\int_s^\beta\left(Q_2^{-1}(t)z_3(t),z_3(t)\right)_n\mathrm{d}t-\left(\overline{B}_1z_4(T),Q_1^{-1}\overline{B}_1z_4(T)\right)_{n-m}-\left(z_3(s),p_2(s)\right)_n.
\end{aligned}\tag{0.54}$$

Relationships (0.52)-(0.54) yield

$$\sigma^2=\left(a,p_2(s)\right)_n,$$

which is to be proved.

It is easy to see that function $z(t)$ defined by (0.45) and the function

$$p(t)=\begin{cases}p_1(t), & 0\le t\le\alpha;\\ p_2(t), & \alpha<t\le s;\\ p_3(t), & s<t\le\beta;\\ p_4(t), & \beta<t\le T,\end{cases}\tag{0.55}$$

satisfy the following uniquely solvable BVP

$$\begin{aligned}
&L^*z(t)=\chi_{(\alpha,\beta)}(t)H^{\mathrm{T}}(t)Q(t)H(t)p_2(t),\quad 0<t<T,\\
&\hat{B}_0z(0)=0,\quad [z]\big|_{t=s}=-a,\quad \hat{B}_1z(T)=0,\\
&Lp(t)=Q_2^{-1}(t)z_1(t),\quad 0<t<T,\\
&B_0p(0)=Q_0^{-1}\overline{B}_0z(0),\quad B_1p(T)=-Q_1^{-1}\overline{B}_1z(T),
\end{aligned}\tag{0.56}$$

where

$$[z]\big|_{t=s} = z(s+0) - z(s-0),$$

$$p \in \left(H^1(0,T)\right)^n,$$

$$z\big|_{(0,s)} \in \left(H^1(0,s)\right)^n, \quad z\big|_{(s,T)} \in \left(H^1(s,T)\right)^n,$$

and $\chi(t)$ is the characteristic function of interval (α,β).

Now Theorem 4.1 can be restated as follows:

Theorem 4.1' The minimax estimate of expression $(a,\varphi(s))_n$ has the form

$$\widehat{\left(a,\varphi(s)\right)_n} = \int_\beta^\alpha \left(\hat{u}(t), y(t)\right)_l \mathrm{d}t + \hat{c},$$

where

$$\hat{u}(t) = Q(t)H(t)p(t), \quad (0.57)$$

$$\hat{c} = \left(\bar{B}_0 z(0), f_0^{(0)}\right)_m - \left(\bar{B}_1 z(T), f_1^{(0)}\right)_{n-m} + \int_0^T \left(z(t), f^{(0)}(t)\right)_n \mathrm{d}t,$$

and vector-functions $z(t)$ and $p(t)$ are determined from the solution to problem 0.56.

Obtain now another representation for the minimax mean square estimate of quantity $(a,\varphi(s))_n$ which is independent of a and s. To this end, introduce vector-functions $\hat{p}_1, \hat{\varphi}_1 \in \left(H^1(0,\alpha)\right)^n$, $\hat{p}_2, \hat{\varphi}_2 \in \left(H^1(\alpha,s)\right)^n$, $\hat{p}_3, \hat{\varphi}_3 \in \left(H^1(s,\beta)\right)^n$, and $\hat{p}_4, \hat{\varphi}_4 \in \left(H^1(\beta,T)\right)^n$ as solutions to the problem

$$L^*\hat{p}_1(t)=0,\quad 0<t<\alpha,\quad \hat{B}_0\hat{p}_1(0)=0,$$
$$L^*\hat{p}_2(t)=H^{\mathrm{T}}(t)Q(t)\left(y(t)-H(t)\hat{\varphi}_2(t)\right),\quad \alpha<t<s,\ \hat{p}_2(\alpha)=\hat{p}_1(\alpha),$$
$$L^*\hat{p}_3(t)=H^{\mathrm{T}}(t)Q(t)\left(y(t)-H(t)\hat{\varphi}_3(t)\right),\quad s<t<\beta,\ \hat{p}_3(s)=\hat{p}_2(s),$$
$$L^*\hat{p}_4(t)=0,\quad \beta<t<T,\quad \hat{p}_4(\beta)=\hat{p}_3(\beta),\quad \hat{B}_1\hat{p}_4(T)=0,$$
$$L\hat{\varphi}_1(t)=Q_2^{-1}(t)\hat{p}_1(t)+f^{(0)}(t),\quad 0<t<\alpha,\ B_0\hat{\varphi}_1(0)=Q_0^{-1}\overline{B}_0\hat{p}_1(0)+f_0^{(0)},$$
$$L\hat{\varphi}_2(t)=Q_2^{-1}(t)\hat{p}_2(t)+f^{(0)}(t),\quad \alpha<t<s,\ \hat{\varphi}_2(\alpha)=\hat{\varphi}_1(\alpha),$$
$$L\hat{\varphi}_3(t)=Q_2^{-1}(t)\hat{p}_3(t)+f^{(0)}(t),\quad s<t<\beta,\ \hat{\varphi}_3(s)=\hat{\varphi}_2(s),$$
$$L\hat{\varphi}_4(t)=Q_2^{-1}(t)\hat{p}_4(t)+f^{(0)}(t),\quad \beta<t<T,$$
$$\hat{\varphi}_4(\beta)=\hat{\varphi}_3(\beta),\quad B_1\hat{\varphi}_4(T)=-Q_1^{-1}\overline{B}_1\hat{p}_4(T)-f_1^{(0)}$$

(0.58)

at realizations y that belong with probability 1 to space $\left(L^2(\alpha,\beta)\right)'$.

Note that unique solvability of problem (0.58) at every realization can be proved similarly to the case of (0.46). Namely, one can show that solutions to the problem of optimal control of the system

$$L^*\hat{p}_1(t;v)=0,\quad 0<t<\alpha,\quad \hat{B}_0\hat{p}_1(0;v)=0,$$

$$L^*\hat{p}_2(t;v)=d(t)-H^{\mathrm{T}}(t)v_1(t),\quad \alpha<t<s,$$

$$\hat{p}_1(\alpha;v)=\hat{p}_2(\alpha;v),$$

$$L^*\hat{p}_3(t;v)=d(t)-H^{\mathrm{T}}(t)v_2(t),\quad s<t<\beta,$$

$$\hat{p}_2(s;v)=\hat{p}_3(s;v),$$

$$L^*\hat{p}_4(t;v)=0,\quad \beta<t<T,$$

$$\hat{p}_4(\beta;v)=\hat{p}_3(\beta;v),\quad \hat{B}_1\hat{p}_4(T;v)=0$$

with the cost function

$$\begin{aligned}J(v)=&\left(Q_0^{-1}\left(\bar{B}_0\hat{p}_1(0;v)+Q_0f_0^{(0)}\right),\bar{B}_0\hat{p}_1(0;v)+Q_0f_0^{(0)}\right)_m\\&+\left(Q_1^{-1}\left(\bar{B}_1\hat{p}_4(T;v)+Q_1f_1^{(0)}\right),\bar{B}_1\hat{p}_4(T;v)+Q_1f_1^{(0)}\right)_{n-m}\\&+\int_0^\alpha\left(Q_2^{-1}(t)\left(\hat{p}_1(t;v)+Q_2(t)f^{(0)}(t)\right),\hat{p}_1(t;v)+Q_2(t)f^{(0)}(t)\right)_n dt\\&+\int_\alpha^s\left(Q_2^{-1}(t)\left(\hat{p}_2(t;v)+Q_2(t)f^{(0)}(t)\right),\hat{p}_2(t;v)+Q_2(t)f^{(0)}(t)\right)_n dt\\&+\int_s^\beta\left(Q_2^{-1}(t)\left(\hat{p}_3(t;v)+Q_2(t)f^{(0)}(t)\right),\hat{p}_3(t;v)+Q_2(t)f^{(0)}(t)\right)_n dt\\&+\int_\beta^T\left(Q_2^{-1}(t)\left(\hat{p}_4(t;v)+Q_2(t)f^{(0)}(t)\right),\hat{p}_4(t;v)+Q_2(t)f^{(0)}(t)\right)_n dt\\&+\int_\alpha^s\left(Q^{-1}(t)v_1(t),v_1(t)\right)_l dt+\int_s^\beta\left(Q^{-1}(t)v_2(t),v_2(t)\right)_l dt\\&\to\min_{v=(v_1,v_2)\in H},\quad d(t)=H^{\mathrm{T}}(t)Q(t)y(t),\alpha<t<\beta,\end{aligned}$$

can be reduced to the solution of problem (0.58) where the optimal control $\hat{v}=(\hat{v}_1,\hat{v}_2)$ is expressed in terms of the solution to this problem as $\hat{v}_1=Q(t)H(t)\hat{\varphi}_2(t),\ \hat{v}_2=Q(t)H(t)\hat{\varphi}_3(t)$; the unique solvability of the problem follows from the existence of the unique minimum point $\hat{v}$ of functional $J(v)$.

Considering system (0.58) at realizations y it is easy to see that its solution is continuous with respect to the right-hand side. This property enables us to conclude, using the general theory of linear continuous transformations of random processes, that the functions $\hat{p}_i(t),\ \hat{\varphi}_i(t)$, $i=\overline{1,4}$, considered as random fields have finite second moments.

Theorem 4.2 The following representation is valid

$$\widehat{\widehat{(a,\varphi(s))_n}}=(a,\hat{\varphi}_2(s))_n.$$

Proof. By virtue of (0.44) and (0.58),

$$\widetilde{\overline{(a,\varphi(s))}}_n = \int_\alpha^s (\hat{u}_1(t), y(t))_l \,\mathrm{d}t + \int_s^\beta (\hat{u}_2(t), y(t))_l \,\mathrm{d}t + \hat{c}$$
$$= \int_\alpha^s (Q(t)H(t)p_2(t), y(t))_l \,\mathrm{d}t + \int_\alpha^s (Q(t)H(t)p_3(t), y(t))_l \,\mathrm{d}t + \hat{c}$$
$$= \int_\alpha^s (p_2(t), H^{\mathrm{T}}(t)Q(t)y(t))_n \,\mathrm{d}t + \int_\alpha^s (p_3(t), H^{\mathrm{T}}(t)Q(t)y(t))_n \,\mathrm{d}t + \hat{c}. \tag{0.59}$$

Next,

$$\int_\alpha^s (p_2(t), H^{\mathrm{T}}(t)Q(t)y(t))_l \,\mathrm{d}t$$
$$= \int_\alpha^s (p_2(t), L^* \hat{p}_2(t))_n \,\mathrm{d}t + \int_\alpha^s (p_2(t), H^{\mathrm{T}}(t)Q(t)H(t)\hat{\varphi}_2(t))_n \,\mathrm{d}t$$
$$= \int_\alpha^s (Lp_2(t), \hat{p}_2(t))_n \,\mathrm{d}t + (p_2(\alpha), \hat{p}_2(\alpha))_n - (p_2(s), \hat{p}_2(s))_n$$
$$+ \int_\alpha^s (H^{\mathrm{T}}(t)Q(t)H(t)p_2(t), \hat{\varphi}_2(t))_n \,\mathrm{d}t$$
$$= \int_\alpha^s (Q_2^{-1}(t)z_2(t), \hat{p}_2(t))_n \,\mathrm{d}t + (p_2(\alpha), \hat{p}_2(\alpha))_n$$
$$- (p_2(s), \hat{p}_2(s))_n - \int_\alpha^s (L^* z_2(t), \hat{\varphi}_2(t))_n \,\mathrm{d}t$$
$$= \int_\alpha^s (z_2(t), L\hat{\varphi}_2(t))_n \,\mathrm{d}t + (p_2(\alpha), \hat{p}_2(\alpha))_n - (p_2(s), \hat{p}_2(s))_n$$
$$- \int_\alpha^s (z_2(t), L\hat{\varphi}_2(t))_n \,\mathrm{d}t + (z_2(s), \hat{\varphi}_2(s))_n$$
$$- (z_2(\alpha), \hat{\varphi}_2(\alpha))_n - \int_\alpha^s (z_2(t), f^{(0)}(t))_n \,\mathrm{d}t$$
$$= (p_2(\alpha), \hat{p}_2(\alpha))_n - (p_2(s), \hat{p}_2(s))_n + (z_2(s), \hat{\varphi}_2(s))_n$$
$$- (z_2(\alpha), \hat{\varphi}_2(\alpha))_n - \int_\alpha^s (z_2(t), f^{(0)}(t))_n \,\mathrm{d}t. \tag{0.60}$$

Similarly,

$$\int_s^\beta (p_3(t), H^{\mathrm{T}}(t)Q(t)y(t))_l \,\mathrm{d}t$$
$$= (p_3(s), \hat{p}_3(s))_n - (p_3(\beta), \hat{p}_3(\beta))_n + (z_3(\beta), \hat{\varphi}_3(\beta))_n$$
$$- (z_3(s), \hat{\varphi}_3(s))_n - \int_s^\beta (z_3(t), f^{(0)}(t))_n \,\mathrm{d}t \tag{0.61}$$

From (0.59), (0.60), and (0.61) it follows

$$\widetilde{\overline{(a,\varphi(s))}} = (p_1(\alpha), \hat{p}_1(\alpha))_n - \int_\alpha^s (z_2(t), f^{(0)}(t))_n \,\mathrm{d}t - \int_s^\beta (z_3(t), f^{(0)}(t))_n \,\mathrm{d}t$$
$$- (z_1(\alpha), \hat{\varphi}_1(\alpha))_n - (p_4(\beta), \hat{p}_4(\beta))_n + (z_4(\beta), \hat{\varphi}_4(\beta))_n + (a, \hat{\varphi}_2(s))_n + \hat{c}. \tag{0.62}$$

However,

$$0=\int_0^\alpha\left(\hat{\varphi}_1(t),L^*z_1(t)\right)_n\mathrm{d}t=\int_0^\alpha\left(L\hat{\varphi}_1(t),z_1(t)\right)_n\mathrm{d}t-\left(z_1(\alpha),\hat{\varphi}_1(\alpha)\right)_n+\left(z_1(0),\hat{\varphi}_1(0)\right)_n$$
$$=\int_0^\alpha\left(z_1(t),Q_2(t)\hat{p}_1(t)\right)_n\mathrm{d}t+\int_0^\alpha\left(z_1(t),f^{(0)}(t)\right)_n\mathrm{d}t-\left(z_1(\alpha),\hat{\varphi}_1(\alpha)\right)_n+\left(z_1(0),\hat{\varphi}_1(0)\right)_n,$$

(0.63)

$$0=\int_0^\alpha\left(L^*\hat{p}_1(t),p_1(t)\right)_n\mathrm{d}t=\int_0^\alpha\left(\hat{p}_1(t),Lp_1(t)\right)_n\mathrm{d}t-\left(\hat{p}_1(\alpha),p_1(\alpha)\right)_n+\left(\hat{p}_1(0),p_1(0)\right)_n$$
$$=\int_0^\alpha\left(\hat{p}_1(t),Q_2(t)z_1(t)\right)_n\mathrm{d}t-\left(\hat{p}_1(\alpha),p_1(\alpha)\right)_n+\left(\hat{p}_1(0),p_1(0)\right)_n.$$

(0.64)

Subtracting from (0.63) equality (0.64), we obtain

$$0=\left(\hat{p}_1(\alpha),p_1(\alpha)\right)_n-\left(z_1(\alpha),\hat{\varphi}_1(\alpha)\right)_n-\left(\hat{p}_1(0),p_1(0)\right)_n+\left(z_1(0),\hat{\varphi}_1(0)\right)_n+\int_0^\alpha\left(z_1(t),f^{(0)}(t)\right)_n\mathrm{d}t$$

or

$$\int_0^\alpha\left(z_1(t),f^{(0)}(t)\right)_n\mathrm{d}t+\left(\hat{p}_1(\alpha),p_1(\alpha)\right)_n-\left(z_1(\alpha),\hat{\varphi}_1(\alpha)\right)_n=\left(\hat{p}_1(0),p_1(0)\right)_n-\left(z_1(0),\hat{\varphi}_1(0)\right)_n.$$

(0.65)

Since

$$\left(z_1(0),\hat{\varphi}_1(0)\right)_n=\left(\bar{B}_0z_1(0),B_0\hat{\varphi}_1(0)\right)_m=\left(\bar{B}_0z_1(0),Q_0^{-1}\bar{B}_0\hat{p}_1(0)\right)_m+\left(\bar{B}_0z_1(0),f_0^{(0)}\right)_m,$$

$$\left(\hat{p}_1(0),p_1(0)\right)_n=\left(\bar{B}_0\hat{p}_1(0),B_0p_1(0)\right)_m=\left(\bar{B}_0\hat{p}_1(0),Q_0^{-1}\bar{B}_0z_1(0)\right)_m,$$

we can use the latter equalities, (0.65), and the fact that Q_0^{-1} is a symmetric matrix to obtain

$$\left(\hat{p}_1(\alpha),p_1(\alpha)\right)_n-\left(z_1(\alpha),\hat{\varphi}_1(\alpha)\right)_n=-\int_0^\alpha\left(z_1(t),f^{(0)}(t)\right)_n\mathrm{d}t-\left(\bar{B}_0z_1(0),f_0^{(0)}\right)_m.$$

(0.66)

Performing a similar analysis, one can prove that

$$-\left(p_4(\beta),\hat{p}_4(\beta)\right)_n+\left(z_4(\beta),\hat{\varphi}_4(\beta)\right)_n=-\int_\beta^T\left(z_4(t),f^{(0)}(t)\right)_n\,\mathrm{d}t+\left(\overline{B}_1 z_1(T),f_1^{(0)}\right)_{n-m}. \tag{0.67}$$

From (0.66), (0.67), and (0.62) and the expression for $\hat{c}$, it follows

$$\widehat{\overline{\left(a,\varphi(s)\right)_n}}=\left(a,\hat{\varphi}_2(s)\right)_n.$$

The theorem is proved.

As is easily seen from (0.58), the functions $\hat{p}\in\left(H^1(0,T)\right)^n$ and $\hat{\varphi}\in\left(H^1(0,T)\right)^n$ defined by

$$\hat{\varphi}(t):=\begin{cases}\hat{\varphi}_1(t), & 0\le t\le\alpha;\\ \hat{\varphi}_2(t), & \alpha<t\le s;\\ \hat{\varphi}_3(t), & s<t\le\beta;\\ \hat{\varphi}_4(t), & \beta<t\le T;\end{cases}$$

and

$$\hat{p}(t):=\begin{cases}\hat{p}_1(t), & 0\le t\le\alpha;\\ \hat{p}_2(t), & \alpha<t\le s;\\ \hat{p}_3(t), & s<t\le\beta;\\ \hat{p}_4(t), & \beta<t\le T;\end{cases},$$

satisfy the following uniquely solvable BVP:

$$\begin{aligned}
&L^{*}\hat{p}(t)=\chi_{(\alpha,\beta)}(t)H^{\mathrm{T}}(t)Q(t)\big(y(t)-H(t)\hat{\varphi}(t)\big),\ 0<t<T,\\
&\hat{B}_0\hat{p}(0)=0,\quad \hat{B}_1\hat{p}(T)=0,\\
&L\hat{\varphi}(t)=Q_2^{-1}(t)\hat{p}(t)+f^{(0)}(t),\quad 0<t<T,\\
&B_0\hat{\varphi}(0)=Q_0^{-1}\overline{B}_0\hat{p}(0)+f_0^{(0)},\\
&B_1\hat{\varphi}(T)=-Q_1^{-1}\overline{B}_1\hat{p}(T)-f_1^{(0)}
\end{aligned} \tag{0.68}$$

at realizations y that belong with probability 1 to space $\left(L^2(\alpha,\beta)\right)^{l}$.

Thus, Theorem 4.2 can be restated in the following form.

Theorem 4.2' The minimax mean square estimate of expression $(a,\varphi(s))_n$ has the form

$$\widehat{\widehat{(a,\varphi(s))_n}}=(a,\hat{\varphi}(s))_n,$$

where vector-function $\hat{\varphi}(t)$ is determined from the solution to problem (0.68).

Remark. Function $\hat{\varphi}(t)$ can be taken as a good, in certain sense, estimate of solution $\varphi(t)$ of initial BVP (0.1), (0.2) on $[0,T]$.

As an example, consider the case when a vector-function $y(t)=H(t)\varphi(t)+\xi(t)$ is observed on an interval $(0,T)$, where a vector-function $\varphi(t)$ with values in $\mathbb{R}^n$ is a solution to the BVP

$$L_1\varphi=f(t),\quad \varphi(0)=f_0,\quad \varphi(T)=f_1, \tag{0.69}$$

and operator L_1 is defined by the relation

$$L_1\varphi(t) = -\varphi''(t) + q(t)\varphi(t),$$

where $q(t)$ is a positive definite $n\times n$-matrix whose entries are continuous functions on $[0,T]$.

Note that this problem has the unique classical solution if $f(t)$ is continuous on $[0,T]$ and the unique generalized solution if $f\in(L_2(0,T))^n$.

Assume that, as well as in the previous case, $H(t)$ is an $l\times n$ matrix with the entries that are continuous functions on $[\alpha,\beta]$ and $\xi(t)$ is a random vector process with zero expectation $\mathbb{E}\xi(t)$ and unknown $l\times l$ correlation matrix $R(t,s)=\mathbb{E}\xi(t)\xi^{\mathrm{T}}(s)$. Assume also that domains V and G are given in the form (0.25) and (0.26) where matrices Q_0, Q_1, and $Q_2(t)$ entering (0.26) have dimensions $n\times n$, $f_0^{(0)}=0,\ f_1^{(0)}=0$, and $f^{(0)}(t)=0.$

Write Equation (0.69) as a first-order system by setting $\varphi_1(t)=\varphi'(t)$, $\varphi_2(t)=\varphi(t)$ and introducing a vector-function

$$\tilde{\varphi}(t)=\begin{pmatrix}\varphi_1(t)\\ \varphi_2(t)\end{pmatrix},\quad \frac{\mathrm{d}\tilde{\varphi}(t)}{\mathrm{d}t}=\begin{pmatrix}\dfrac{\mathrm{d}\varphi_1(t)}{\mathrm{d}t}\\ \dfrac{\mathrm{d}\varphi_2(t)}{\mathrm{d}t}\end{pmatrix},$$

$$\tilde{f}(t)=\begin{pmatrix}-f(t)\\ 0\end{pmatrix},\quad \tilde{f}_0=\begin{pmatrix}0\\ f_0\end{pmatrix},\quad \tilde{f}_1=\begin{pmatrix}0\\ f_1\end{pmatrix}$$

with $2n$ components, a vector $\tilde{a}=(0,a)$ with $2n$ components, a $2n\times 2n$-matrix

$$\tilde{A}(t)=\begin{pmatrix} O_{n,n} & -q(t) \\ -E_n & O_{n,n} \end{pmatrix},$$

matrices $B_0=B_1=\bar{B}_0=\bar{B}_1=(O_{n,n},E_n)$, and $\hat{B}_0=\hat{B}_1=(E_n,O_{n,n})$. Then system (0.69) can be written as

$$\tilde{L}\tilde{\varphi}(t):=\frac{\mathrm{d}\tilde{\varphi}(t)}{\mathrm{d}t}+\tilde{A}\tilde{\varphi}(t)=\tilde{f}(t),\quad B_0\tilde{\varphi}(0)=\tilde{f}_0, \tag{0.70}$$

$$B_1\tilde{\varphi}(T)=\tilde{f}_1. \tag{0.71}$$

Applying Theorems 1 and 2 and performing necessary transformations in the resulting equations that are similar to (0.46) and (0.58) (in terms of the designations introduced above) we prove the following Theorem 4.3 The minimax mean square estimate of expression $(a,\varphi(s))_n$ has the form

$$\widehat{\widehat{(a,\varphi(s))_n}}=\int_0^s(\hat{u}_1(t),y(t))_l\,\mathrm{d}t+\int_s^T(\hat{u}_2(t),y(t))_l\,\mathrm{d}t=(a,\hat{\varphi}_2(s))_n,$$

where

$$\sigma^2=(a,p_2(s))_n,$$

$\hat{u}_2(t)=Q(t)H(t)p_3(t),\ \hat{u}_1(t)=Q(t)H(t)p_2(t),$ and vector-functions $p_2(t)$, $p_3(t)$ and $\hat{\varphi}_2(t)$ are determined from the solutions to the problems

$$L_1z_2(t)=-H^{\mathrm{T}}(t)Q(t)H(t)p_2(t),\ 0<t<s,\ z_2(0)=0,$$

$$L_1z_3(t)=-H^{\mathrm{T}}(t)Q(t)H(t)p_3(t),\quad s<t<T,$$

$$z_2(s) - z_3(s) = a, \quad z_3(T) = 0,$$

$$L_1 p_2(t) = Q_2^{-1}(t) z_2(t), \ 0 < t < s, \ p_2(0) = Q_0^{-1} z_2(0),$$

$$L_1 p_3(t) = Q_2^{-1}(t) z_3(t), \quad s < t < T,$$

$$p_2(s) = p_3(s), \quad p_3(T) = -Q_1^{-1} z_3(T),$$

and

$$L_1 \hat{p}_2(t) = H^{\mathrm{T}}(t) Q(t) \big(y(t) - H(t) \hat{\varphi}_2(t) \big), \quad 0 < t < s,$$

$$\hat{p}_2(0) = 0,$$

$$L_1 \hat{p}_3(t) = H^{\mathrm{T}}(t) Q(t) \big(y(t) - H(t) \hat{\varphi}_3(t) \big), \quad s < t < T,$$

$$\hat{p}_3(s) = \hat{p}_2(s), \quad \hat{p}_3(T) = 0,$$

$$L_1 \hat{\varphi}_2(t) = Q_2^{-1}(t) \hat{p}_2(t), \ 0 < t < s, \ \ \hat{\varphi}_2(0) = Q_0^{-1} \hat{p}_2(0),$$

$$L_1 \hat{\varphi}_3(t) = Q_2^{-1}(t) \hat{p}_3(t), \quad s < t < T,$$

$$\hat{\varphi}_3(s) = \hat{\varphi}_2(s), \quad \hat{\varphi}_3(T) = -Q_1^{-1} \hat{p}_3(T),$$

respectively.

5. MINIMAX MEAN SQUARE ESTIMATES OF SOLUTIONS SUBJECT TO INCOMPLETE RESTRICTIONS ON UNKNOWN PARAMETERS

Assume again that observations have form (0.24) and unknown parameters f_0, f_1, and $f(t)$ belong to the domain

$$G=\left\{\tilde{F}=\left(\tilde{f}_0,\tilde{f}_1,\tilde{f}\right):\int_0^T\left(Q_2(t)\tilde{f}(t),\tilde{f}(t)\right)dt\le 1\right\}, \tag{0.72}$$

where $Q_2(t)$ is given in (0.26). The correlation function of process $\xi(t)$ belongs to domain (0.25). Introduce the set

$$U=\left\{u(\cdot):\overline{B}_0 z_1(0,u)=0,\overline{B}_1 z_4(T,u)=0\right\} \tag{0.73}$$

here $u(t)=\begin{cases}u_1(t), & \alpha<t<s,\\ u_2(t), & s<t<\beta,\end{cases}$ where $z_i(t,u),\ i=\overline{1,4},$ is the solution to BVP (0.29).

Lemma 5.1

$$\sigma(u,c)=\begin{cases}\infty, & u\notin U,\\ \sigma_1(u,c), & u\in U,\end{cases}$$

where

$$\sigma_1(u,c)=\int_0^T\left(Q_2^{-1}\tilde{z}(t,u),\tilde{z}(t,u)\right)dt+\int_\alpha^\beta\left(Q^{-1}(t)u(t),u(t)\right)dt+c^2=J(u)+c^2. \tag{0.74}$$

This lemma can be proved using formula (0.39).

Lemma 5.2 U is a convex closed set in the space $\left(L_2(\alpha,\beta)\right)'$.

Proof. The convexity of set U is obvious. Let us prove that this set is closed.

Note that functions $z_1(0,u)$ and $z_4(T,u)$ can be represented as

$$z_1(0,u) = a_1 + \int_\alpha^\beta \Phi_1(t)u(t)\mathrm{d}t,$$
$$z_4(T,u) = a_2 + \int_\alpha^\beta \Phi_2(t)u(t)\mathrm{d}t, \tag{0.75}$$

where $\Phi_1(t)$ and $\Phi_2(t)$ are known matrix functions with the elements from $L_2(\alpha,\beta)$ and a_1 and a_2 are vectors. Expression (0.75) can be obtained if we introduce a vector z_0 such that $z_1(0,u) = z_0$. Then $z_1(\alpha,u) = \Phi(\alpha,0)z_0$,

$$\begin{aligned} z_2(s,u) &= \Phi(s,\alpha)z_2(\alpha,u) \\ &= \Phi(s,\alpha)z_1(\alpha,u) + \int_\alpha^s \Phi(s,\tau)H^{\mathrm{T}}(\tau)u_1(\tau)\mathrm{d}\tau \\ &= \Phi(s,0)z_0 + \int_\alpha^s \Phi(s,\tau)H^{\mathrm{T}}(\tau)u_1(\tau)\mathrm{d}\tau, \end{aligned}$$

where $\Phi(t,\tau)$ is a solution to the equation

$$\frac{\mathrm{d}\Phi(t,\tau)}{\mathrm{d}t} = -A^*(t)\Phi(t,\tau), \quad \Phi(\tau,\tau) = E$$

and E is the unit matrix. Next,

$$z_4(T,u) = \Phi(T,0)z_0 + \int_\alpha^\beta \Phi(T,\tau)H^{\mathrm{T}}(\tau)u(\tau)\mathrm{d}\tau + \Phi(T,s)a.$$

Since BVP (0.29) is uniquely solvable, there exists one and only one vector z_0 satisfying the system of algebraic equations

$$\begin{cases} \bar{B}_0 z_0 = 0, \\ B_1 \Phi(T,0) z_0 = -\int_\alpha^\beta B_1 \Phi(t,\tau) H^{\mathrm{T}}(\tau) u(\tau) \mathrm{d}\tau - B_1 \Phi(T,s) a. \end{cases}$$

Solving this system we determine z_0 in the form

$$z_0 = b + \int_\alpha^\beta \Phi_0(\tau) u(\tau) \mathrm{d}\tau,$$

where $\Phi_0(\tau)$ is a known matrix function continuous on $[\alpha,\beta]$ and b is a known vector. Taking into account this equality, we obtain expressions (0.75). From these relationships, it follows that if a sequence $u_n(t)$ converges in $L_2(\alpha,\beta)$ to a function $u_0(t)$, then

$$\lim_{n\to\infty} \bar{B}_0 z_1(0,u_n) = \bar{B}_0 z_1(0,u_0),$$

$$\lim_{n\to\infty} \bar{B}_1 z_4(0,u_n) = \bar{B}_1 z_4(0,u_0),$$

which proves that U is a closed set.

Assume now that U is nonempty. Then the following statement is valid.

Theorem 5.1 There exists the unique minimax mean square estimate of expression $(a,\varphi(s))$ which can be represented in the form (0.44) at $\hat{c} = 0$, where vector-functions $p_2(t)$ and $p_3(t)$ solve the equations

$$
\begin{aligned}
&L^{*}z_1(t)=0,\quad 0<t<\alpha,\quad \hat{B}_0 z_1(0)=0,\\
&L^{*}z_2(t)=-H^{\mathrm{T}}(t)Q(t)H(t)p_2(t),\quad \alpha<t<s,\ z_2(\alpha)=z_1(\alpha),\\
&L^{*}z_3(t)=-H^{\mathrm{T}}(t)Q(t)H(t)p_3(t),\quad s<t<\beta,\ z_3(s)=z_2(s)-a,\\
&L^{*}z_4(t)=0,\quad \beta<t<T,\quad z_4(\beta)=z_3(\beta),\\
&\hat{B}_1 z_4(T)=0,\quad \bar{B}_0 z_1(0),\quad \bar{B}_1 z_4(T)=0,\\
&Lp_1(t)=Q_2^{-1}(t)z_1(t),\quad 0<t<\alpha,\\
&Lp_2(t)=Q_2^{-1}tz_2(t),\quad \alpha<t<s,\quad p_2(\alpha)=p_1(\alpha),\\
&Lp_3(t)=Q_2^{-1}(t)z_3(t),\quad s<t<\beta,\quad p_3(s)=p_2(s),\\
&Lp_4(t)=Q_2^{-1}(t)z_4(t),\quad s<t<\beta.
\end{aligned}
\tag{0.76}
$$

Proof. Similarly to Theorem 4.1 one can show that for $u\in U$ the following equality holds

$$\sigma(u,c)=J(u)+c^2,$$

where

$$
\begin{aligned}
J(u)=&\int_0^{\alpha}\left(Q_2^{-1}(t)z_1(t,u),z_1(t,u)\right)_n \mathrm{d}t+\int_{\alpha}^{s}\left(Q_2^{-1}(t)z_2(t,u),z_2(t,u)\right)_n \mathrm{d}t\\
&+\int_s^{\beta}\left(Q_2^{-1}(t)z_3(t,u),z_3(t,u)\right)_n \mathrm{d}t+\int_{\beta}^{T}\left(Q_2^{-1}(t)z_4(t,u),z_4(t,u)\right)_n \mathrm{d}t\\
&+\int_{\alpha}^{s}\left(Q^{-1}(t)u_1(t),u_1(t)\right)\mathrm{d}t+\int_s^{\beta}\left(Q^{-1}(t)u_2(t),u_2(t)\right)\mathrm{d}t,
\end{aligned}
$$

and $z_i(t,u), i=\overline{1,4},$ are solutions to Equations (0.29) at $\bar{B}_0 z_1(0,u)=0$ and $\bar{B}_1 z_4(T,u)=0.$ $J(u)$ is a strictly convex lower semicontinuous functional on a closed convex set U and $\lim_{\|u\|\to+\infty} J(u)=\infty.$ Therefore there exists one and only one vector $\hat{u}$ such that $\min_{u\in U} J(u)=J(\hat{u}).$ This vector can be determined from the relationship

$$\frac{\mathrm{d}}{\mathrm{d}\tau}J_{\mu}\left(\hat{u}_1+\tau v_1,\hat{u}_2+\tau v_2\right)\Big|_{\tau=0}\equiv 0,\quad \forall v=(v_1,v_2)\in H,$$

where

$$J_\mu(u)=J(u_1,u_2)+(\mu_1,\bar{B}_0 z_1(0,u))+(\mu_2,\bar{B}_1 z_4(T,u)),$$

$\mu_1\in\mathbb{R}^m$, $\mu=(\mu_1,\mu_2)$, and $\mu_2\in\mathbb{R}^{n-m}$ are Lagrange multipliers.

Further analysis is similar to the proof of Theorem 4.1. Let vector-functions $\hat{p}_i(t)$ and $\hat{\varphi}_i(t)$, $i=\overline{1,4}$, be solutions to the system

$$
\begin{aligned}
&L^*\hat{p}_1(t)=0,\quad 0<t<\alpha,\\
&L^*\hat{p}_2(t)=H^{\mathrm{T}}(t)Q(t)[y(t)-H(t)\hat{\varphi}_2(t)],\quad \alpha<t<s,\\
&L^*\hat{p}_3(t)=H^{\mathrm{T}}(t)Q(t)[y(t)-H(t)\hat{\varphi}_3(t)],\quad s<t<\beta,\\
&L^*\hat{p}_4(t)=0,\quad \beta<t<T,\\
&L\hat{\varphi}_1(t)=Q_2^{-1}(t)\hat{p}_1(t),\quad 0<t<\alpha,\\
&L\hat{\varphi}_2(t)=Q_2^{-1}(t)\hat{p}_2(t),\quad \alpha<t<s,\\
&L\hat{\varphi}_3(t)=Q_2^{-1}(t)\hat{p}_3(t),\quad s<t<\beta,\\
&L\hat{\varphi}_4(t)=Q_2^{-1}(t)\hat{p}_4(t),\quad \beta<t<T,\\
&\hat{B}_0\hat{p}_1(0)=0,\quad \hat{p}_2(\alpha)=\hat{p}_1(\alpha),\\
&\hat{p}_3(s)=\hat{p}_2(s),\quad \hat{p}_4(\beta)=\hat{p}_3(\beta),\\
&\hat{B}_1\hat{p}_1(T)=0,\quad \bar{B}_0\hat{p}_1(0)=0,\quad \bar{B}_1\hat{p}_4(T)=0,\\
&\hat{\varphi}_2(\alpha)=\hat{\varphi}_1(\alpha),\quad \hat{\varphi}_3(s)=\hat{\varphi}_2(s),\quad \hat{\varphi}_4(\beta)=\hat{\varphi}_3(\beta).
\end{aligned}
\tag{0.77}
$$

Theorem 5.2 Assume that for any vector $a\in\mathbb{R}^n$ set U is nonempty. Then system (0.77) is uniquely solvable and the equality

$$\widehat{\widehat{(a,\varphi(s))}}=(a,\hat{\varphi}_2(s))$$

holds Proof. Introduce functions $\hat{p}_i(t,v)$, $i=\overline{1,4}$, as a solution to the BVP

$$L^*\hat{p}_1(t,v)=0,\quad 0<t<\alpha,$$

$$L^{*}\hat{p}_2(t,v)=d(t)-H^{T}(t)v_1(t), \quad \alpha<t<s,$$

$$L^{*}\hat{p}_3(t,v)=d(t)-H^{T}(t)v_2(t), \quad s<t<\beta,$$

$$L^{*}\hat{p}_4(t,v)=0, \quad \beta<t<T,$$

$$\hat{B}_0\hat{p}_1(0,v)=0, \quad \hat{B}_1\overline{p}_4(T,v)=0,$$

$$\hat{p}_2(\alpha,v)=\hat{p}_1(\alpha,v), \quad \hat{p}_3(s,v)=\hat{p}_2(s,v),$$

$$\hat{p}_4(\beta,v)=\hat{p}_3(\beta,v),$$

where $d(t)=H^{T}(t)Q(t)y(t)$. Define a set

$$U_1=\{v:\overline{B}_0\hat{p}_1(0,v)=0,\overline{B}_1\hat{p}_4(T,v)=0\}.$$

Since U is nonempty, the same is valid for U_1 for any vector a. Similarly to the case of U one can show that U_1 is a convex closed set. Denote by $J_1(v)$ the functional of the form

$$\begin{aligned}J_1(v)=&\int_0^{\alpha}\left(Q_2^{-1}(t)\hat{p}_1(t,v),\hat{p}_1(t,v)\right)dt+\int_{\alpha}^{s}\left(Q_2^{-1}(t)\hat{p}_2(t,v),\hat{p}_2(t,v)\right)dt\\&+\int_s^{\beta}\left(Q_2^{-1}(t)\hat{p}_3(t,v),\hat{p}_3(t,v)\right)dt+\int_{\beta}^{T}\left(Q_2^{-1}(t)\hat{p}_4(t,v),\hat{p}_4(t,v)\right)dt\\&+\int_{\alpha}^{s}\left(Q^{-1}(t)v_1(t),v_1(t)\right)dt+\int_s^{\beta}\left(Q^{-1}(t)v_2(t),v_2(t)\right)dt.\end{aligned}$$

One can show, following Theorem 5.1, that on set U_1 there is one and only one point of minimum of functional $J_1(v)$, namely,

$$\hat{v}_1(t)=Q^{-1}(t)H(t)\hat{\varphi}_2(t), \quad \hat{v}_2(t)=Q^{-1}(t)H(t)\hat{\varphi}_3(t),$$

where functions $\hat{\varphi}_2(t)$ and $\hat{\varphi}_3(t)$ are determined from system (0.77). The proof of the equality

$$\widehat{\widehat{(a,\varphi(s))_n}} = (a,\hat{\varphi}_2(s))_n$$

is similar to that in Theorem 4.2.

6. CONCLUSIONS

For a system described by a one-dimensional two-point BVP with decoupling boundary conditions at the endpoints of the interval and quadratic restrictions imposed on the unknown deterministic data and the second moments of observation noise, we have obtained guaranteed mean square estimates of inner product $(a,\varphi(s))$, where $\varphi(s)$ is the unknown solution of the BVP at a point $s\in(0,T)$ and $a\in\mathbb{R}^n$. Guaranteed estimates are obtained using the duality of the problems of estimation and optimal control. We have shown that guaranteed mean square estimates and estimation errors are expressed via solutions to special optimal control problems for conjugate BVPs. The solutions to these optimal control problems enable one to find explicit expressions for estimates and estimation errors both for distributed and point observations.

The obtained results are applied to minimax estimation of solutions of two-point BVPs for linear ordinary second-order differential equations.

Methods and results of the paper may be used for estimation under uncertainties of the states of the systems described by more general linear BVPs for different classes of functional--differential equations; in particular, for systems of differential equations with impulse perturbations, differential equations with multipoint conditions, and in several other cases.

7. RESULTS OF NUMERICAL EXPERIMENTS

Let realizations of the random variables

$$y_k = \int_0^1 g_k(x)\varphi(x)\mathrm{d}x + \xi_k, \quad k = 1,\cdots,N, \tag{0.78}$$

be observed. Here ξ_k are independent random variables for which $\mathbb{E}\xi_k = 0, \mathbb{E}\xi_k^2 = r_k^2$; $\varphi(x)$ is a solution of the BVP

$$-\varphi''(x) + \omega^2\varphi(x) = f(x), \tag{0.79}$$

$$\varphi(0) = \varphi(1) = 0; \tag{0.80}$$

and

$$g_k(x) = \sqrt{2}\sin k\pi x, \quad k = 1,2,\cdots$$

are eigenfunctions of the operator $-\frac{\mathrm{d}^2}{\mathrm{d}x^2} : H_0^2(0,1) \to L^2(0,1)$, where $H_0^2(0,1) = \{u \in H^2(0,1) : u(0) = u(1) = 0\}$.

We assume that function $f(x)$ is not known exactly and is chosen arbitrarily from the set $G = \left\{f(x) : \int_0^1 f^2(x)\mathrm{d}x \le q^2\right\}$, where q is a certain constant.

Applying the technique similar to the estimation methods developed in Section 4, it is possible to obtain expressions for the minimax mean square estimates in the case when observations have the form (0.78). In particular, in this case the function $\hat{\varphi}$ which approximates the solution φ of BVP (0.79)-(0.80) on the interval $[0,1]$ is determined from the system

$$-\hat{p}''(x)+\omega^2\hat{p}(x)=\sum_{k=1}^{N} r_k^{-2} g_k(x)\left(y_k-(g_k,\hat{\varphi})_{L^2(0,1)}\right),$$

$$\hat{p}(0)=\hat{p}(1)=0,$$

$$-\hat{\varphi}''(x)+\omega^2\hat{\varphi}(x)=q^2\hat{p}(x),$$

$$\hat{\varphi}(0)=\hat{\varphi}(1)=0,$$

and can be represented in the form

$$\hat{\varphi}(x)=\sum_{k=1}^{N}\left(\frac{r_k^2\left((k\pi)^2+\omega^2\right)^2}{q^2}+1\right)^{-1} g_k(x)\,y_k.$$

The exact solution φ and its estimate $\hat{\varphi}$ (bold curves) calculated on the basis of the modelled observations are presented in Figure 1 and Figure 2. Calculations are performed at $q=1$, $w=1$, and $N=5$ for $k=1,\cdots,5$ using the parameters $r_k=r=\frac{1}{100}$ and $f(x)=\frac{1}{2000}\sin\pi x+\frac{1}{2}\sin 2\pi x+\frac{1}{2}\sin 3\pi x+\sin 4\pi x$ (Figure 1) and $r_k=r=\frac{1}{500}$ and $f(x)=\frac{1}{2}\cos\left(10x+\frac{1}{2}\right)$ (Figure 2).

As can be seen from these figures, parameter r plays the crucial role as far as the estimation quality is concerned. In fact, this parameter directly influences the signal-to-noise ratio.

The calculations were performed using Wolfram Mathematica.

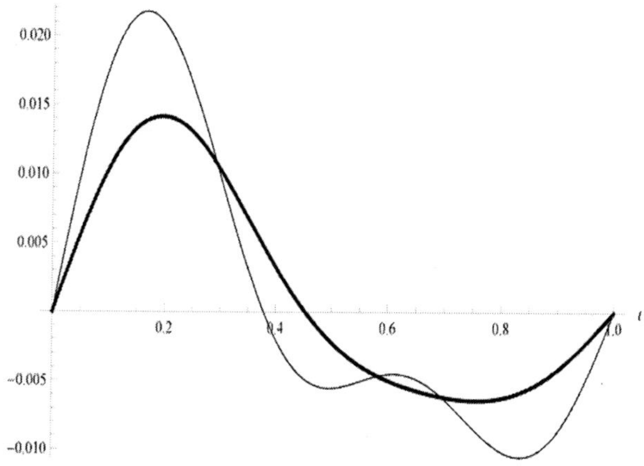

Figure 1. Exact solution φ and its estimate $\hat{\varphi}$ (bold curve) calculated at q = 1, w = 1, N = 5, $r_k = r = \frac{1}{100}(k = 1,\ldots,5)$, and $f(x) = \frac{1}{2000}\sin \pi x + \frac{1}{2}\sin 2\pi x + \frac{1}{2}\sin 3\pi x + \sin 4\pi x$.

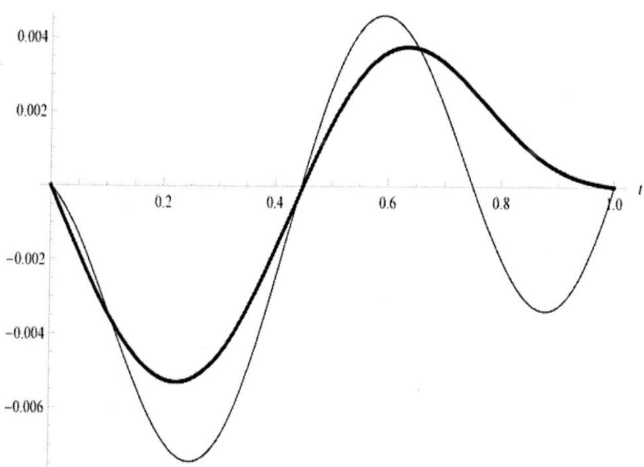

Figure 2. Exact solution φ and its estimate $\hat{\varphi}$ (bold curve) calculated at q = 1, w = 1, N = 5, $r_k = r = \frac{1}{500}(k = 1,\ldots,5)$, and $f(x) = \frac{1}{2}\cos\left(10x + \frac{1}{2}\right)$.

REFERENCES

1. Kalman, R. (1960) A New Approach to Liner Filtering and Prediction Problems. Journal of Basic Engineering, 82, 35-45.

2. Kalman, R. and Bucy, R. (1961) New Results in Liner Filtering and Prediction Theory. Transactions of the ASME-Journal of Basic Engineering, 83, 95-108.

3. Krasovskii, N. (1968) Theory of Motion Control. Nauka, Moscow.

4. Kurzhanskii, A. (1977) Control and Observation under Uncertainties. Nauka, Moscow.

5. Kirichenko, N. and Nakonechnyi, O. (1977) A Minimax Approach to Recurrent Estimation of the States of Linear Dynamical Systems. Kibernetika, 4, 52-55.

6. Nakonechnyi, O. (1979) Minimax Estimates in Systems with Distributed Parameters. Preprint 79, Acad. Sci. USSR, Inst. Cybernetics, Kyiv, 55 p.

7. Kuntsevich, V. (2005) Accuracy of Construction of Approximating Models under Bounded Measurement Noises. Automation and Remote Control, 66, 791-798.

8. Kurzhanski, A. and Valyi, I. (1997) Ellipsoidal Calculus for Estimation and Control. Birkhauser Verlag, Basel.

9. Nakonechnyi, O., Podlipenko, Y. and Shestopalov, Y. (2009) Estimation of Parameters of Boundary Value Problems for Linear Ordinary Differential Equations with Uncertain Data. arXiv:0912.2872v1, 1-72.

10. Podlipenko, Y. (2005) Minimax Estimation of Right-Hand Sides of Noetherian Equations in a Hilbert Space under Uncertainty Conditions. Reports of the National Academy of Sciences of Ukraine, 12, 36-44.

11. Basar, T. and Bernhard, P. (1991) H-Optimal Control and Related Minimax Design Problems. Birkhäuser, Basel.

12. Fedoryuk, M. (1985) Ordinary Differential Equations. Nauka, Moscow.

13. Naimark, M. (1969) Linear Differential Operators. Nauka, Moscow.

14. Krein, S. (1971) Linear Equations in the Banach Space. Nauka, Moscow.

15. Atkinson, F. (1964) Discrete and Coninuous Boundary Value Problems. Academic Press, New York.

16. Hutson, V., Pym, J. and Cloud, M. (2005) Applications of Functional Analysis and Operator Theory. Vol. 200, 2nd Edition, Mathematics in Science and Engineering, Elsevier Science, Amsterdam.

8

BOUNDARY VALUE PROBLEMS FOR BURGERS EQUATIONS, THROUGH NONSTANDARD ANALYSIS

Saida Bendaas

Department of Mathematics, Faculty of Sciences,
University of Setif 1, Elbez, Algeria

ABSTRACT

In this paper we study inviscid and viscid Burgers equations with initial conditions in the half plane $x \in R, T > 0$. First we consider the Burgers equations with initial conditions admitting two and three shocks and use the HOPF-COLE transformation to linearize the problems and explicitly solve them. Next we study the Burgers equation and solve the initial value problem for it. We study the asymptotic behavior of solutions and we show that the exact solution of boundary value problem for viscid Burgers equation as viscosity parameter is sufficiently small approach the shock type solution of boundary value problem for inviscid Burgers equation. We discuss both confluence and interacting shocks. In this article a new approach has been developed to find the exact solutions. The results are formulated in classical mathematics and proved with infinitesimal technique of non standard analysis.

KEYWORDS

Non Standard Analysis, Boundary Value Problem, Viscid Burgers Equation, Inviscid Burgers Equation, Heat Equation

1. INTRODUCTION

The nonlinear parabolic partial differential equation

$$u_t + uu_x = \varepsilon u_{xx} \tag{1.1}$$

was first introduced by J. M. Burgers [1] [2] as the simplest model for fluid flow, this equation combining both nonlinear propagation effects and diffusive effect. If ε is not null, we approach to the Naveir's Stokes equations in one dimension. Burgers equation has a wide variety of applications in the modeling of water in unsaturated soil, dynamics of soil water, statistics of flow problems mixing and turbulent diffusion cosmology and seismology.

When ε is null, this equation approaches to the Euler's equations in one dimension who governs the flows of perfect fluids. It's the viscid equation. it has the form

$$u_t + uu_x = 0. \tag{1.2}$$

If the viscous term is dropped from the Burgers equation, discontinuities may appear in finite time; even if the initial condition is smooth, they give rise to the phenomenon of shock waves with important application in physics [3] . This property makes Burgers equation a proper model for testing numerical algorithms in flows where severe gradients or shocks are anticipated [4] -[6] . Discretization methods are well-known techniques for solving Burgers equation. Ascher and McLachlan established many methods as multi-symplectic box sheme. For the boundary value problem, Sinai [7] was interested to the initial condition case: null on R_- and Brownian on R_+. She, Aurell and Frich [8] with numerical calculations particularly examined the initial conditions of Brownian fraction nair to the asymptotic behavior.

A remarkable feature of viscid Burgers equation is that its solutions with initial conditions of the form

$$u(\xi,0)=f(\xi) \tag{1.3}$$

can be explicitly written down. Hopf [9] and Cole [10] independently showed that the Equation (1.1) can be linearized through the transformation

$$u=-2\varepsilon\varphi_x/\varphi. \tag{1.4}$$

Then Hopf [9] showed that if $\phi(x,t)$ satisfies the linear heat equation

$$\varphi_t=\varepsilon\varphi_{xx} \tag{1.5}$$

with initial condition

$$\varphi(x,0)=\exp\left[-\frac{1}{2\varepsilon}\left[\int_0^\infty f(v)\,\mathrm{d}v+\frac{(x-\eta)^2}{2t}\right]\right]\mathrm{d}\eta. \tag{1.6}$$

Then $u(x,t)$ defined by Equation (1.4) solves (1.1) and (1.3). Conversely, if $u(x,t)$ is a solution of problem (1.1) and (1.3) then $\varphi(x,t)$ defined by Equation (1.4) is a solution of problem (1.5) and (1.6). Solving for φ from (1.5) and (1.6) and substituting it into Equation (1.3) we obtained explicit formula for the solution of problem (1.1) and (1.2) namely:

$$u(x,t)=\frac{\int_{-\infty}^{\infty}\frac{x-\eta}{t}\exp\left[-\frac{1}{2\varepsilon}\left[\int_0^\infty f(v)\,\mathrm{d}v+\frac{(x-\eta)^2}{2t}\right]\right]\mathrm{d}\eta}{\int_{-\infty}^{\infty}\exp\left[-\frac{1}{2\varepsilon}\left[f(v)\,\mathrm{d}v+\frac{(x-\eta)^2}{2t}\right]\right]\mathrm{d}\eta} \tag{1.7}$$

and studied the asymptotic behavior of $u(x,t)$.

Explicit solutions of the Burgers equation (1.1) in the quarter plane with integrable initial data and piecewise constant boundary data were constructed by [11] using Hopf-Cole transformation [9] . He obtained a formula for its weak limit as viscosity parameter goes to 0. Although, there are many results for initial value problem has been studied less. Using maximum principle, this formula for weak limit was extended to general boundary data. ε is a positive parameter small enough. The problem is considered by [12] . As the fact that ε multiplies the largest derivative, one is in the presence of a singular perturbation problem. The purpose of Singular Perturbation Theory is to investigate the behavior of solutions of (1.1) as $\varepsilon \to 0$.

The aim of the present article is to study solutions of Inviscid and Viscid Burgers equation if the initial condition admits several singular points, i.e. in the case of a finite number of shocks. A simple formulation is given for the asymptotic behavior based on the evaluation of integrals which is a method of the non standard perturbation theory of differential equations proposed by Imm Van Den Berg [13] and improved by Lutz and Callot.

Historically the subject non standard was developed by Robinson, Reeb, Lutz and Goze [14]. The nonstan- dard perturbation theory of differential equations, which is today a well-established tool in asymptotic theory, has its roots in the seventies, when the Reebian school (see [14] [15]) introduced the use of nonstandard analysis into the field of perturbed differential equations. Our goal in this paper is to generalize these techniques on EDP and our general purpose is to describe the asymptotic behavior of solutions in boundary value problem with a small parameter ε and to discuss in particular the cases of confluence and the interacting shocks with new technical infinitesimal of non-standard analysis. We can conclude that the solutions of the problem: (1.1) and (1.3) are infinitely close to the solutions of problems (1.2) and (1.3), as ε is a parameter positive sufficiently small.

In Section 2, we treat the boundary value problem for inviscid Burgers equation, solve it and study it. Section 3 is devoted to useful lemmas for our main results. In Section 4, we study viscid Burgers equation, solve exactly the initial value problems for it, and describe the asymptotic behavior of solutions with a non standard form. Some components, such as multi-leveled equations, graphics, and tables are not prescribed, although the various table text styles are provided. The formatter will need to create these components, incorporating the applicable criteria that follow.

2. INITIAL BOUNDARY VALUE PROBLEM FOR INVISCID BURGERS' EQUATION

2.1. Shock Fitting

We consider the inviscid Burgers equation:

$$u_t + uu_x = 0.$$

In: $x \in R$, $t > 0$ with the initial condition

$$u(\xi,0) = f(\xi)$$

where $f : R \to R,$ is continuous.

This problem not admits the regular solutions but some weak solutions with certain regularity exist. The Burgers equation on the whole line is known to possess traveling wave solutions. The solution of (1.2) and (1.3) may be given in a parametric form and shocks must be fitted in such that:

$$U = \frac{1}{2}(u_1 + u_2) = \frac{1}{2}\left(f(\xi_1) + f(\xi_2)\right) \tag{2.1}$$

where ξ_1 and ξ_2 are the value of ξ on the two sides of the shock [16] .

According to Equation (1.2), the solution at time t is obtained from the initial profile $u = f(\xi)$ by translating each point a distance $f(\xi)t$ to the right. The shock cuts out the part corresponding to $\xi_2 \geq \xi \geq \xi_1$. If the discontinuity line, it is a straight line chord property still holds. The cord on the f curve cuts off lobes of equal area. The shock determination can then be describe entirely on the fixe $f(\xi)$ curve by drawing all the chords with the equal area property can be written analytically as between the points $\xi = \xi_1$ and $\xi = \xi_2$ on the curve $f(\xi)$. Moreover since areas are preserved under the mapping, the equal area

$$\frac{1}{2}(f(\xi_1)+f(\xi_2))(\xi_1-\xi_2)=\int_{\xi_1}^{\xi_2} f(\xi)\mathrm{d}\xi. \quad (2.2)$$

This is the differential equation for the line cord of shock that checks the condition of entropy such as [16] [17] Corresponding to the two inflection points.

2.2. Confluence of Shocks

When a number of shocks are produced, in general it is possible for one of them to overtake the shock ahead. Then they combine and continue as a single shock. This is also described by our shock solution.

Consider the curve given by f Figure 1, then two shocks are formed corresponding to the inflection points p and q with families of equal area chords, typified by p_1p_2 and q_1q_2.

As time goes, the points q_1 and p_2 approach each other until the stage in Figure 2 is reached where a common chord cuts off lobes of equal area for both humps.

At this stage the characteristics corresponding to p_2' and q_1' are the same and therefore the shocks have just combined into one as shown in Figure 3. Characteristics p_2' and q_1' combined. All the characteristics between q_2' and p_1' have now been absorbed by one or other of the shocks. A single shock proceeds using chords $p_1''q_2''$.

In the plane (x,t) the shocks can be represented by Figure 4.

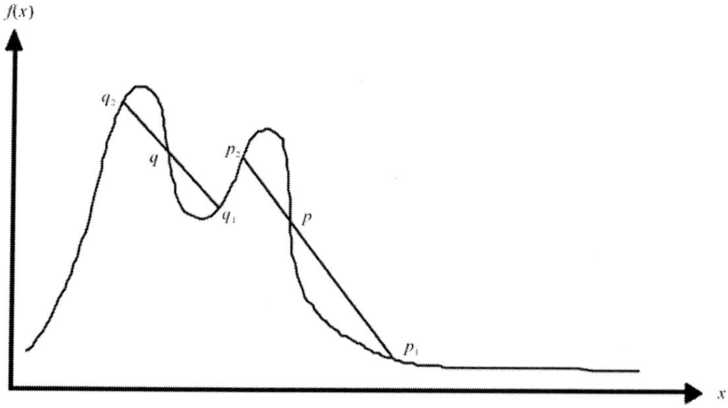

Figure 1. Graphic representation of the initial condition f.

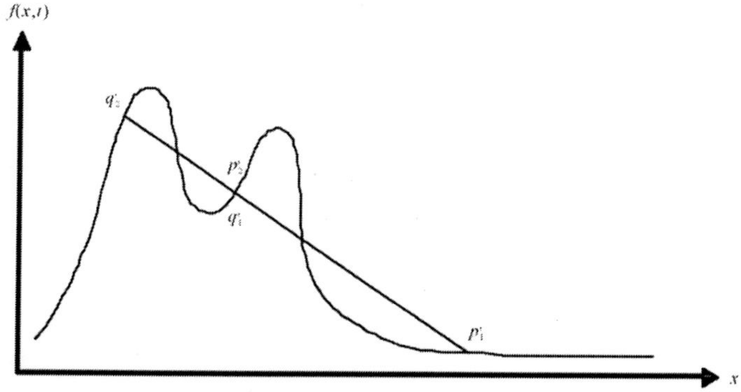

Figure 2. Graphical representation from the merger of the two shocks. The characteristics p_2' and q_1' merge.

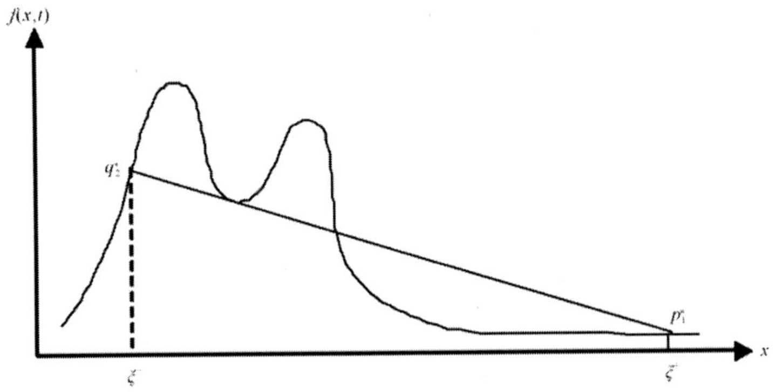

Figure 3. Construction for merging shocks in a final stage.

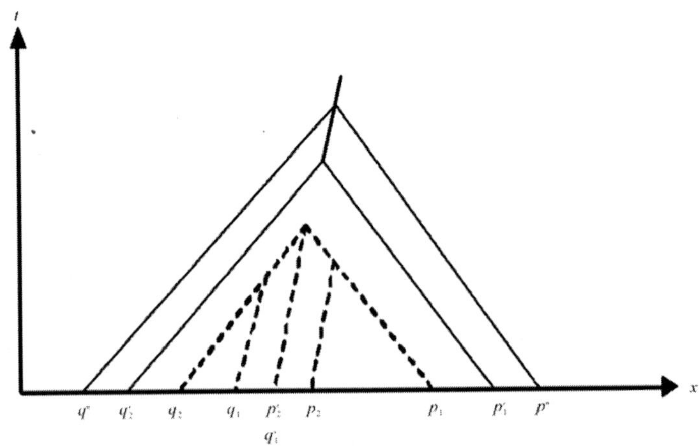

Figure 4. The (x, t) diagram for merging shocks corresponding to Figure 1.

3. PRELIMINARIES

In this section we present some lemmas that are important to prove our main result.

Proposition 3.1. Let $\varphi = \varphi_1 + \varphi_2$ be the analytic solution for the initial value problem for the heat equation (1.5) and (1.6). If the initial condition is given as in Figure 1, then u given by the Formula (1.4) is a solution for the initial value problem (1.1) and (1.3). It is explicitly given by:

$$u = -2\varepsilon \frac{\phi_x}{\phi} = \frac{u_1\varphi_1 + u_2\varphi_2}{\varphi_1 + \varphi_2} \tag{3.1}$$

i.e.

$$u(x,t) = \frac{\sum_{i=1}^{i=2} \int_{-\infty}^{\infty} \frac{x-\eta_i}{t} \exp\left[-\frac{1}{2\varepsilon} \int_0^{\eta_i} f(v)\,\mathrm{d}v + \frac{(x-\eta_i)^2}{2t} \right] \mathrm{d}\eta}{\sum_{i=1}^{i=2} \frac{1}{\sqrt{4\pi\varepsilon}} \int_{-\infty}^{\infty} \exp\left[-\frac{1}{2\varepsilon} \int_0^{\eta_i} f(v)\,\mathrm{d}v + \frac{(x-\eta_i)^2}{2t} \right] \mathrm{d}\eta}. \tag{3.2}$$

Proof: When a shock overtakes another shock, they merge into a single shock of increased strength as described in inviscid solution $(\varepsilon = 0)$, on the f curve in Figure 1; It is possible to give a simple solution of Burgers equation that describes this process for arbitrary ε. The solution for a single shock is given in [3] [4] and the corresponding expression for φ may be written in the form:

$$\varphi = \varphi_1 + \varphi_2; \ \varphi_i = \exp\left(\frac{u_i}{2\varepsilon} x + \frac{u_i^2}{4\varepsilon} t - b_i \right), \ i = 1,2. \tag{3.3}$$

In the expression of solution for a single shock given in [3] [4] , the parameters b_1, b_2 witch locate the initial position of the shock are taken to be zero. The expressions $\varphi_1 \varphi_2$ are clearly solutions of the heat equation $(\varphi_t = \varepsilon\varphi_{xx})$ with

$$\varphi_i = \exp\left(\frac{u_i}{2\varepsilon} x + \frac{u_i^2}{4\varepsilon} t \right), i = 1,2. \tag{3.4}$$

Corresponding to the initial conditions:

$$\varphi_0 = \phi_0(\eta_i) = \exp\left\{-\frac{1}{2\varepsilon}\int_0^x f(\eta_i)\mathrm{d}\eta_i\right\}, i=1,2 \tag{3.5}$$

Then the solutions of the heat equation are given as:

$$\varphi_i = \frac{1}{\sqrt{4\pi\varepsilon t}}\exp\left[-\frac{1}{2\varepsilon}\int_0^{\eta_i} f(\nu)\mathrm{d}\nu + \frac{(x-\eta_i)^2}{2t}\right]\mathrm{d}\eta_i, i=1,2 \tag{3.6}$$

$$u_i\varphi_i = \int_{-\infty}^{+\infty}\frac{x-\eta_i}{t}\exp\left[-\frac{1}{2\varepsilon}\int_0^{\eta_i} f(\nu)\mathrm{d}\nu + \frac{(x-\eta_i)^2}{2t}\right]\mathrm{d}\eta_i \tag{3.7}$$

Using Equations (1.4) and (3.1) we obtain the expression (3.2).

And to prove our results, we use the non standard analysis techniques, for that we consider the following lemma.

Lemma 3.2. (The Van. Den. Berg lemma [14]): Let h be a standard function, defined and increasing on [0,+∞[such that $h(\nu) = a\nu^r(1+\delta)$ where δ for v ≃ 0. $h(\nu) > m(\nu)^q$. And let φ be an intern function defined on]0,+∞[such that : $\varphi(\nu) = b\nu^s(1+\delta)$ for v ≈ 0, and such that ∀ d > 0, ∃ standard k and standard c

such that: $|\varphi(\nu)| < k\exp(\cosh(\nu))$, for $\nu > d$. Then

$$\int_0^\infty \varphi(\nu)\exp\left(\frac{-h(\nu)}{2\varepsilon}\right)\mathrm{d}\nu = \frac{b\Gamma\left(\frac{s+1}{r}\right)}{ra^{\left(\frac{s+1}{r}\right)}}\cdot\frac{1}{\left(\frac{1}{2\varepsilon}\right)^{\left(\frac{s+1}{r}\right)}} \tag{3.8}$$

where a, r are positive standard, m and q are the both positive δ is an infinitesimal. b and s are standard, $b \neq 0$ and $s > -1$.

To give estimation to the solution, given by (3.2), we state the following lemma:

Lemma 3.3. Let ε be a positive real small enough. And let ϕ and h be two standard functions such that: h, is a C^2 class function verified the Lemma 3, and admits on the ξ point a unique absolute minimum $h'(\xi)=0$ and $h''(\xi)>0$. $\varphi(\xi)\neq 0$. It is S-continuous on ξ and satisfies the conditions of the Lemma 3 in the both ways. Then

$$\int_{-\infty}^{\infty}\varphi(\eta)\exp\left(\frac{-h(\eta)}{2\varepsilon}\right)\mathrm{d}\eta=\varphi(\xi)\frac{\sqrt{4\pi\varepsilon}}{\sqrt{h''(\xi)}}(1+\delta) \tag{3.9}$$

δ is an infinitesimal.

Proof: To prove this lemma, we use the "Van Den Berg" method, lemmas: (5.6), (5.7) [14] . It consists in the following steps

1) Search for the absolute minimum (maximum) of the function under the exponential sign and bring it out.

2) Bring back the minimum (maximum) to the zero.

3) Searching the galaxy as well as the main galaxy where the function in the exponential sign is appreciable.

4) Calculate the integral.

As consequence we have the following lemma.

Lemma 3.4. Let f the initial condition as Figure 1. Assume:

(H_1): $f:R\rightarrow R$ is $C^2(R)$.

(H_2): There exist a, b, c, d and e in R, with a < b < c < d < e, such that

$$\begin{cases} f''(x)\geq 0, & \text{if } x\in]-\infty,a[,]b,c[\text{ and if } x\in]c,d[,]e,\infty[\\ f''(x)\leq 0, & \text{if } x\in[a,b] \text{ and if } x\in[d,e] \end{cases}$$

Then for x and t fixed, the functions defined as:

$$h_i(\eta_i, x, t) = \int_0^{\eta} f(v)\mathrm{d}v + \frac{(x-\eta)^2}{2t}, i = 1,2 \tag{3.10}$$

has at most two minima ξ_i^+ and ξ_i^- relative to the variable η_i. These two minima satisfy the equations:

$$x = \eta_i + tf(\eta_i), i = 1,2 \tag{3.11}$$

And the condition: $h_i(\xi_i^-, x, t) = h_i(\xi_i^+, x, t)$, is equivalent to the shock conditions:

$$\frac{1}{2}\left(f(\xi_i^-) + f(\xi_i^+)\right)\left(\xi_i^- - \xi_i^+\right) = \int_{\xi_i^-}^{\xi_i^+} f(v)\mathrm{d}v, i = 1,2 \tag{3.12}$$

Proof: Let f the initial condition given as in Figure 1, two shocks are formed corresponding to the inflection points of $f(\xi)$. for some time we see appear an area of three values for the solution $u(x,t)$. At first the wave breaks on the feature for which: $f_i'(x) < 0$ and $|f_i'(x)|$ is a maximum during the time $t = t_c$. Inside the zone of each shock and for a point (x,t), there are three characteristics corresponding to two minimum framing a maximum. When a shock overtakes another, they merge into a single shock of increased strength as describe for the inviscid solution $(\varepsilon = 0)$ on the f curve in Figure 2. The characteristics between q_2' and p_1' are absorbed by oneFigure 3. At this stage there are two stationary values that satisfy the equations (3.11) we noted them by ξ_i^- and $\xi_i^+, i = 1,2$, each couple frames a maximum. Let: $\eta_i(x,t)$, be the minimum for the functions given by the Formula (3.10), is such that:

$$\frac{\partial h_i}{\partial \eta_i} = f(\eta_i) - \frac{x - \eta_i}{t} = 0, i = 1,2$$

This equation is verified at the two minima ξ_i^- and $\xi_i^+, i=1,2$.

If $h_i(\xi_i^-,x,t)=h_i(\xi_i^+,x,t)$, and within (3.10) this condition can be written as:

$$\int_0^{\xi_i^+} f(v)\mathrm{d}v+\frac{(x-\xi_i^+)^2}{2t}=\int_0^{\xi_i^-} f(v)\mathrm{d}v+\frac{(x-\xi_i^-)^2}{2t}, i=1.2 \tag{3.13}$$

But ξ_i^- and ξ_i^+ both verify the equations:

$$f(\xi_i)-\left(\frac{x-\xi_i}{t}\right)=0, i=1,2 \tag{3.14}$$

The condition of the shock is expressed by (3.12), is the same condition of shock given by (2.2) for invicid Burgers equation.

4. INITIAL BOUNDARY VALUE PROBLEM FOR INVISCID BURGERS' EQUATION

4.1. Confluence of Shocks

Our general purpose now is to show that the exact solution of (1.1) and (1.3) endorse the ideas regarding shocks in Section 2, we want to confirm that as ε is small enough, the solution of (1.1) and (1.3) reduce to solution of (1.2) and (1.3), with discontinuous shocks which satisfy the condition (2.2), and the shocks are located at the positions determined in Section 2. The shocks are formed corresponding to the inflection points of the initial condition $u(x,0)=f(x)$, who assume the assumptions (H_1), (H_2) in the lemma (3.4). Then we proved the following result:

Theorem 4.1. Under the assumptions: (H_1), (H_2) in lemma (3.4), the problem (1.1) and (1.3) admits a unique solution for $t>0$ given by:

$$u(x,t)=\frac{\sum_{i=1}^{2}\int_{-\infty}^{+\infty}\frac{x-\eta_i}{t}\exp\left[-\frac{1}{2\varepsilon}\int_0^{\eta_i}f(v)\mathrm{d}v+\frac{(x-\eta_i)^2}{2t}\right]\mathrm{d}\eta_i}{\sum_{i=1}^{2}\frac{1}{\sqrt{4\pi\varepsilon}}\int_{-\infty}^{\infty}\exp\left[-\frac{1}{2\varepsilon}\int_0^{\eta_i}f(v)\mathrm{d}v+\frac{(x-\eta_i)^2}{2t}\right]\mathrm{d}\eta_i} \tag{4.1}$$

Such a solution is confluence of shocks and for ε sufficiently small, this solution is infinitely close to the solution of the reduced problem given in (2.2).

Proof: 1) From Figure 2 and Figure 3, after some time the two shocks combine and continue into one, and this is the lowest minimum that carries this amount to the case of a single shock. From the proposition (3.1), the problem (1.1) and (1.3) admits a single solution explicitly given by Formula (4.1). Uniqueness is due to the condition of entropy which restricts the set of solutions to one which is stable with a singular perturbation dissipative nature.

2) Let (x,t) be a standard point outside the line of the shock. From the solutions of the equation $x=\eta_i+tf(\eta_i)$, there is only one denoted ξ_i is the absolute minimum of the function given by the expression (3.10). Using the expressions (3.1) and (3.2) of the Proposition 3.1 $u(x,t)$ is given as:

$$u(x,t)=\frac{I}{J}=\frac{\sum_{i=1}^{2}\mu_i\varphi_i}{\sum_{i=1}^{2}\varphi_i}=\frac{\sum_{i=1}^{2}\int_{-\infty}^{+\infty}\frac{x-\eta_i}{t}\exp\left[-\frac{1}{2\varepsilon}\int_0^{\eta_i}f(v)\mathrm{d}v+\frac{(x-\eta_i)^2}{2t}\right]\mathrm{d}\eta_i}{\sum_{i=1}^{2}\frac{1}{\sqrt{4\pi\varepsilon}}\int_{-\infty}^{\infty}\exp\left[-\frac{1}{2\varepsilon}\int_0^{\eta_i}f(v)\mathrm{d}v+\frac{(x-\eta_i)^2}{2t}\right]\mathrm{d}\eta_i}$$

From the Lemma 3.2 we have

$$u(x,t)=\frac{I}{J}=\frac{\sum_{i=1}^{2}\mu_i\varphi_i}{\sum_{i=1}^{2}\varphi_i}=\frac{\sum_{i=1}^{2}\frac{x-\eta_i}{t}\frac{\sqrt{4\pi\varepsilon}}{\sqrt{h''(\xi)}}\left[-\frac{h(\eta_i)}{2\varepsilon}\right](1+\delta)}{\sum_{i=1}^{2}\frac{\sqrt{4\pi\varepsilon}}{\sqrt{h''(\xi)}}\left[-\frac{h(\eta_i)}{2\varepsilon}\right](1+\delta)}$$

where δ > 0 is an infinitesimal. And we will have the following estimate

$$u(x,t)=\frac{x-\eta_i}{t}(1+\delta)\simeq u_i(x,t)=f(\xi_i), i=1,2$$

To conclude we have the following corollary.

Corollary 4.2. Let (x,t), be a standard point outside each line of shock. Among the solutions of the equation $x=\eta_i+tf(\eta_i)$, one denoted ξ_i is the absolute minimum of the function $h_i(\eta_i,x,t)$ given by (3.10), and further the solution of (1.1) and (1.3) verifies at the point (x,t)

$$u(x,t)\simeq f(\xi_1)=u_1(x,t), \textit{ if } x \textit{ is an infinitely large positive,}$$

$$u(x,t)\simeq f(\xi_2)=u_2(x,t), \textit{ if } x \textit{ is an infinitely large negative.}$$

And the center of the shock when $\varphi_1=\varphi_1$ is that:

$$x=\frac{1}{2}(u_1+u_2)$$

Proof: Using lemma (3.4), outside the region of each shock. For (x,t) fixed, each function $h_i(\eta_i,x,t)$ has an absolute minimum at $\xi_i, i=1,2$. In (3.1) $u(x,t)$ is in the form

$$u(x,t)=\frac{u_1\varphi_1+u_2\varphi_2}{\varphi_1+\varphi_2}$$

With φ_i and $u_i\varphi_i$ are given by (3.6) and (3.7). Then

$$u(x,t)=\frac{\sum_{i=1}^{2}\int_{-\infty}^{+\infty}\frac{x-\eta_i}{t}\exp\left[-\frac{1}{2\varepsilon}\int_0^{\eta_i}f(v)\mathrm{d}v+\frac{(x-\eta_i)^2}{2t}\right]\mathrm{d}\eta_i}{\sum_{i=1}^{2}\frac{1}{\sqrt{4\pi\varepsilon}}\int_{-\infty}^{\infty}\exp\left[-\frac{1}{2\varepsilon}\int_0^{\eta_i}f(v)\mathrm{d}v+\frac{(x-\eta_i)^2}{2t}\right]\mathrm{d}\eta_i}$$

Using lemma (3.3) we obtain

$$u(x,t)=\frac{\left(\frac{x-\eta_1}{t}\frac{\sqrt{4\pi\varepsilon}}{\sqrt{h''(\xi_1)}}\exp\left[-\frac{h(\eta_1)}{2\varepsilon}\right]+\frac{x-\eta_2}{t}\frac{\sqrt{4\pi\varepsilon}}{\sqrt{h''(\xi_2)}}\exp\left[-\frac{h(\eta_2)}{2\varepsilon}\right]\right)(1+\delta)}{\left(\frac{\sqrt{4\pi\varepsilon}}{\sqrt{h''(\xi_1)}}\exp\left[-\frac{h(\eta_1)}{2\varepsilon}\right]+\frac{\sqrt{4\pi\varepsilon}}{\sqrt{h''(\xi_2)}}\exp\left[-\frac{h(\eta_2)}{2\varepsilon}\right]\right)(1+\delta)}$$

And it follows that:

If $u_2 > u_1$, φ_1 function dominates when x is infinitely large positive and there is obtained

$$u(x,t)=\frac{x-\eta_1}{t}(1+\delta)\simeq u_1(x,t)=f(\xi_1),$$

If $u_1 > u_2$, φ_2 function dominates when x is infinitely large negative and there is obtained

$$u(x,t)=\frac{x-\eta_2}{t}(1+\delta)\simeq u_2(x,t)=f(\xi_2).$$

4.2. Interacting Shocks

In this section, we discuss the interacting shocks case; before going further in this case we need the following proposition and lemma.

Now since any φ_i a solution of the heat equation, we may clearly add further terms in (3.3) and generate more general solution of burgers' equation. Such solution represents interacting shocks. As consequence we have the following

Proposition 4.3. Let $\varphi=\varphi_1+\varphi_2+\varphi_3$, be the analytical solution for the problem (1.5) and (1.3). If f is the initial condition admitting three inflection points, then the solution of (1.1) and (1.3) is explicitly given by

$$u(x,t) = -2\varepsilon\frac{\varphi_x}{\varphi} = \frac{u_1\varphi_1 + u_2\varphi_2 + u_3\varphi_3}{\varphi_1 + \varphi_2 + \varphi_3} \tag{4.2}$$

i.e.

$$u(x,t) = \frac{\sum_{i=1}^{3}\int_{-\infty}^{+\infty}\frac{x-\eta_i}{t}\exp\left[-\frac{1}{2\varepsilon}\int_0^{\eta_i} f(v)\,\mathrm{d}v + \frac{(x-\eta_i)^2}{2t}\right]\mathrm{d}\eta_i}{\sum_{i=1}^{3}\frac{1}{\sqrt{4\pi\varepsilon}}\int_{-\infty}^{\infty}\exp\left[-\frac{1}{2\varepsilon}\int_0^{\eta_i} f(v)\,\mathrm{d}v + \frac{(x-\eta_i)^2}{2t}\right]\mathrm{d}\eta_i}, \quad i = 1,2,3 \tag{4.3}$$

Proof: When a shock overtakes another, they merge into a single shock of increased strength as described in inviscid solution $(\varepsilon = 0)$. For arbitrary ε, it is possible to provide a simple solution to the Burgers equation that describes this process. The solution for a single shock is given in [4] and the corresponding expression for φ is written as:

$$\varphi = \varphi_1 + \varphi_2 + \varphi_3,\ \ \varphi_i = \exp\left(-\frac{u_i}{2\varepsilon}x + \frac{u_i^2}{4\varepsilon}t - b_i\right).$$

In the solution for a single shock given in [4] , the parameters b_1, b_2 which locate the initial position of the

shock are taken to be zero and $b_3 = \frac{1}{2\varepsilon}(u_3 - u_2)$. The expressions φ_1, φ_2 and φ_3 are clearly solution of the

heat equation $(\varphi_t = \varepsilon\varphi_{xx})$ with

$$\varphi_i = \exp\left[-\frac{u_i}{2\varepsilon}x + \frac{u_i^2}{4\varepsilon}t\right],\ i = 1,2. \tag{4.5}$$

And

$$\varphi_3 = \exp\left[-\frac{u_3}{2\varepsilon}x + \frac{u_3^2}{4\varepsilon}t - \frac{u_3 - u_2}{2\varepsilon}\right], i = 3 \tag{4.6}$$

Corresponding to the initial conditions:

$$\varphi_0 = \phi_0(\eta_i) = \exp\left\{-\frac{1}{2\varepsilon}\int_0^x f(\eta_i)\,\mathrm{d}\eta\right\}, i = 1,2,3 \tag{4.7}$$

Then the solutions of the heat equation are given as:

$$\varphi_i = \frac{1}{\sqrt{4\pi\varepsilon t}}\exp\left[-\frac{1}{2\varepsilon}\int_0^{\eta_i} f(v)\,\mathrm{d}v + \frac{(x-\eta_i)^2}{2t}\right]\mathrm{d}\eta_i, i = 1,2,3 \tag{4.8}$$

and

$$u_i\varphi_i = \int_{-\infty}^{+\infty}\frac{x-\eta_i}{t}\exp\left[-\frac{1}{2\varepsilon}\int_0^{\eta_i} f(v)\,\mathrm{d}v + \frac{(x-\eta_i)^2}{2t}\right]\mathrm{d}\eta_i, i = 1,2,3 \tag{4.9}$$

Using (1.4) and (4.4) we obtain the expression (4.3). Then we have the following.

Theorem 4.4. For $t > 0$, and for $u_3 > u_2 > u_1 > 0$, the problems (1.1) and (1.3) have a unique solution given by:

$$u(x,t) = \frac{\sum_{i=1}^{3}\int_{-\infty}^{+\infty}\frac{x-\eta_i}{t}\exp\left[-\frac{1}{2\varepsilon}\int_0^{\eta_i} f(v)\,\mathrm{d}v + \frac{(x-\eta_i)^2}{2t}\right]\mathrm{d}\eta}{\sum_{i=1}^{3}\frac{1}{\sqrt{4\pi\varepsilon}}\int_{-\infty}^{\infty}\exp\left[-\frac{1}{2\varepsilon}\int_0^{\eta_i} f(v)\,\mathrm{d}v + \frac{(x-\eta_i)^2}{2t}\right]\mathrm{d}\eta}$$

Such a solution is interacting shocks and for ε sufficiently small, it is infinitely close to the solution of the reduced problems (1.2) and (1.3).

Proof. 1) In the interacting shock case we have three shocks, when a shock overtakes another they merge into a single shock of increased strength and the lowest minimum dominating. Then we go back to the single shock case. Using the proposition (4.3), we deduce the uniqueness of solution explicitly given by (3.4). The uniqueness is due to the entropic condition [3] [4] , which restricts the set of solutions to one, who is stable with singular perturbation with dissipative nature.

2) Let (x,t) be a standard point outside the line of the shock. From the solutions of the equation $x=\eta_i+tf(\eta_i)$, there is only one denoted that ξ_i is the absolute minimum of the function given by the expression (3.10). $u(x,t)$ is given as

$$u(x,t)=\frac{I}{J}=\frac{\sum_{i=1}^{3}u_i\varphi_i}{\sum_{i=1}^{3}\varphi_i}=\frac{\sum_{i=1}^{3}\int_{-\infty}^{+\infty}\frac{x-\eta_i}{t}\exp\left[-\frac{1}{2\varepsilon}\int_0^{\eta_i}f(v)\mathrm{d}v+\frac{(x-\eta_i)^2}{2t}\right]\mathrm{d}\eta_i}{\sum_{i=1}^{3}\frac{1}{\sqrt{4\pi\varepsilon}}\int_{-\infty}^{\infty}\exp\left[-\frac{1}{2\varepsilon}\int_0^{\eta_i}f(v)\mathrm{d}v+\frac{(x-\eta_i)^2}{2t}\right]\mathrm{d}\eta_i}$$

From the Lemma 3.2 we have:

$$u(x,t)=\frac{I}{J}=\frac{\sum_{i=1}^{3}u_i\varphi_i}{\sum_{i=1}^{3}\varphi_i}=\frac{\sum_{i=1}^{3}\frac{x-\eta_i}{t}\frac{\sqrt{4\pi\varepsilon}}{\sqrt{h''(\xi_i)}}\left[-\frac{h_i(\eta_i)}{2\varepsilon}\right](1+\delta)}{\sum_{i=1}^{3}\frac{\sqrt{4\pi\varepsilon}}{\sqrt{h''(\xi_i)}}\left[-\frac{h_i(\eta_i)}{2\varepsilon}\right](1+\delta)}$$

where $\delta > 0$ is an infinitesimal. And we will have the following estimate

$$u(x,t)=\frac{x-\eta_i}{t}(1+\delta)\simeq u_i(x,t)=f(\xi_i),\ i=1,2,3$$

from which the following corollary

Corollary 4.5. Let (x,t) be a standard point outside the line of the shock. So there exists a unique solution denoted ξ_i which is the absolute minimum of the function given by (3.10). If ε is reasonably small, we can recognise shock transition between the states u_1, u_2, u_3 by noting in which regions the corresponding φ dominates. At $t=0$, φ_1 dominates in $0<x$, φ_2 in: $-1<x<0$; φ_3 dominates in: $x<-1$. Then

$$u(x,t) \simeq u_1(x,t) = f(\xi_1), \text{ in } 0 < x,$$

$$u(x,t) \simeq u_2(x,t) = f(\xi_2), \text{ in } -1 < x < 0,$$

$$u(x,t) \simeq u_3(x,t) = f(\xi_3), \text{ in } x < -1.$$

Proof: For (x,t) fixed, outside the shock region and by the use of the lemma (3.2), each function h_i admits one absolute minimum on the point ξ_i. In Formula (3.4), $u(x,t)$ is as

$$u = \frac{u_1\varphi_1 + u_2\varphi_2 + u_3\varphi_3}{\varphi_1 + \varphi_2 + \varphi_3} \tag{4.11}$$

where $\varphi_i, u_i\varphi_i$ are given by (4.8) and (4.9); and using the Lemma 3.2 there are equivalent to those

$$\varphi_i = \frac{\sqrt{4\pi\varepsilon}}{\sqrt{h''(\xi_i)}} \exp\left[-\frac{h_i(\xi_i)}{2\varepsilon}\right](1+\delta), i = 1,2,3 \tag{4.12}$$

$$u_i\varphi_i = \frac{x-\eta}{t}\frac{\sqrt{4\pi\varepsilon}}{\sqrt{h''(\xi_i)}} \exp\left[-\frac{h_i(\xi_i)}{2\varepsilon}\right](1+\delta), i = 1,2,3 \tag{4.13}$$

where δ is an infinitesimal positive real.

As shown in Figure 5, If ε is reasonably small, we can recognise shock transition between the states u_1, u_2, u_3 by noting in which regions the corresponding φ dominates. At $t=0$, in: $0<x$, φ_1 dominates and we have:

$$u(x,t)\simeq u_1(x,t)=f(\xi_1),$$

φ_2 dominates in: $-1<x<0$, then we have:

$$u(x,t)\simeq u_2(x,t)=f(\xi_2),$$

In: $x<-1$, φ_3 dominates and we obtain:

$$u(x,t)\simeq u_3(x,t)=f(\xi_3),$$

The symbol "$\simeq$" means infinitely close to [8] . Thus we have a shock transitions from u_1 to u_2 centred at $x=0$, and one from u_2 to u_3 centered at $x=-1$. For $t>0$, for early times the transition from u_1 to u_2 occurs where $\varphi_1=\varphi_2$ on

$$x=\frac{1}{2}(u_1+u_2)t \tag{4.14}$$

The transition from u_2 to u_3 occurs where on

$$x=\frac{1}{2}(u_2+u_3)t-1 \tag{4.15}$$

Since, $\frac{1}{2}(u_2+u_3)>\frac{1}{2}(u_1+u_2)$, the second shock overtakes the first at the point (x^*,t^*) determined by (4.14) and (4.15). At this point

$$\varphi_1=\varphi_2=\varphi_3$$

for $t > t^*$, there is no longer any region where φ_2 dominates and the continuing solution describe a single shock transition between u_1 and u_3, moving with the velocity

$$x = \frac{1}{2}(u_1 + u_3)t$$

on the path determined by $\varphi_1 = \varphi_3$.

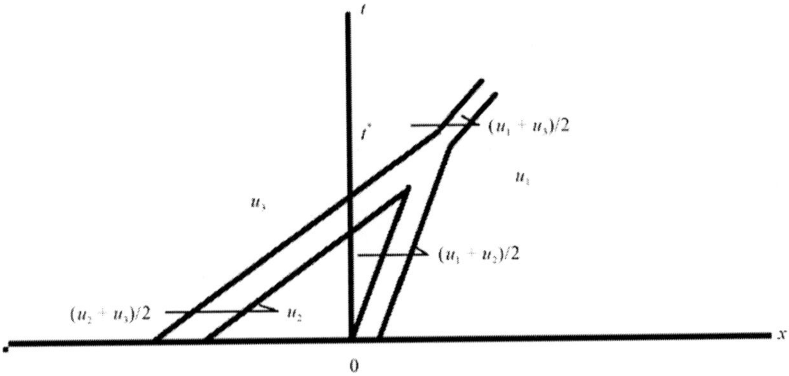

Figure 5. Merging shocks.

ACKNOWLEDGEMENTS

Saida Bendaas I thank the editor and the referee for their comments.

REFERENCES

1. Burgers, J.M. (1948) A Mathematical Model Illustrating the Theory of Turbulence. In: Von Mises, R. and Von Karman, T., Eds., Advances in Applied Mechanics, Vol. 1, Academic Press, New York, 171-

199.http://www.sciencedirect.com/science/article/pii/S00652156 08701005

2. Burgers, J.M. (1975) The Non Linear Diffusion Equation Asymptotic Solution and Statistical Problems. http://www.amazon.it/The-Non-Linear-Diffusion-Equation-Statistical/dp/9027704945

3. Kida, S. (1979) Asymptotic Properties of Burgers Turbulence. Journal of Fluid Mechanics, 93, 337-377.http://journals.cambridge.org/action/displayAbstract;jsessionid=0174C93AF976CBF18771FEB41E0FEFA9.journals?fromPage=online&aid=388434

4. Samokhin, A. (2014) Gradient Catastrophes and Saw Tooth Solution for a Generalized Burgers Equation on an Interval. Journal of Geometry and Physics, 85, 177-184.http://www.sciencedirect.com/science/article/pii/S0393044014000965

5. Samokhin, A.V. (2013) Evolution of Initial Data for Burgers Equation with Fixed Boundary Values. Sci Herald of MSTUCA, 194, 63-70.http://www.mstuca.ru/scientific_work/scientific_work/files/194.pdf

6. Euvrard, D. (1992) Résolution Numérique des Equations aux Dérivées Partielles. Différences finies, Eléments finis. Masson, Paris.

7. Sinai, G. (1992) Statistics of Shocks in Solutions of Inviscid Burgers Equation. Communications in Mathematical Physics, 148, 601-621.http://citeseerx.ist.psu.edu/showciting?cid=205153

8. She, Z.S., Aurell, E. and Frich, U. (1992) The Inviscid Burgers Equation with Initial Data of Brownian Type. Com- munications in Mathematical Physics, 148, 623-641.http://www.researchgate.net/publication/38331252_The_inviscid_Burgers_equation_with_initial_data_of_Brownian_type

9. Hopf, E. (1950) The Partial Differential Equation: $u_t + uu_x = \varepsilon u_{xx}$. Communications on Pure and Applied Mathematics, 3, 201-230.http://www.researchgate.net/publication/259149172_The_partial_differential_equation_ut__uux__xx

10. Cole, J.D. (1951) On a Quasilinear Parabolic Equation Occurring in Aerodynamics. Quarterly of Applied Mathematics, 9, 225-236.http://www.researchgate.net/publication/238286127_On_a_quasilinear_parabolic_equation_occurring_in_aerodynamics

11. Joseph, K.T. (1988) Burgers Equation in the Quarter Plane, a Formula for the Weak Limit. Communications on Pure and Applied Mathematics, 41, 133-149.http://onlinelibrary.wiley.com/ doi/ 10.1002/cpa.3160410202/abstracthttp://dx.doi.org/10.1002/cpa. 3160410202

12. Kevorkian, J. and Cole, J.D. (1981) Perturbation Methods in Applied Mathematics. Springer Verlag, New York. http:// www.amazon.com/Perturbation-Methods-Mathematics-Mathematical-Sciences/dp/0387905073 http://dx.doi.org/ 10. 1007/978-1-4757-4213-8

13. Van Den Berg, I. (1987) Non Standard Asymptotic Analysis. Lecture Notes in Mathematics, 1249.http://www.researchgate.net/ publication/44553479_Nonstandard_asymptotic_analysis__Imme_van_den_Berg

14. Lutz, R. and Goze, M. (1981) Non Standard Analysis. A Practical Guide with Application. Lecture Notes in Mathe- matics, 861.

15. Lutz, R. and Sari, T. (1982) Applications of Nonstandard Analysis in Boundary Value Problems in Singular Perturbatio Theory; Theory and Application of Singular Perturbation (Oberwolfach 1981). Lecture Notes in Mathematics, 942, 113-135.http:// www. researchgate.net/publication/225547873_Applications_of_nonstandard_analysis_to_boundary_value_problems_in_singular_perturbation_theory

16. Bendaas, S. (1994) Quelques applications de l'A.N.S aux E.D.P. Ph.D. Thesis, Haute Alsace University, France.

17. Bendaas, S. (2008) L'équation de Burgers avec un Terme Dissipatif. Une approche non standard. Analele Universitatii din Oradea. Fascicola Matematica, 15, 239-252.http://stiinte.uoradea.ro/en/auofm/auofm_contents.htm

INDEX

O

Q

R

S

T

U

V